William Fawcett

Economic Plants

An index to economic products of the vegetable kingdom in Jamaica

William Fawcett

Economic Plants

An index to economic products of the vegetable kingdom in Jamaica

ISBN/EAN: 9783743407930

Manufactured in Europe, USA, Canada, Australia, Japa

Cover: Foto ©berggeist007 / pixelio.de

Manufactured and distributed by brebook publishing software
(www.brebook.com)

William Fawcett

Economic Plants

ECONOMIC PLANTS.

AN INDEX

TO

ECONOMIC PRODUCTS

OF

THE VEGETABLE KINGDOM

IN

JAMAICA.

COMPILED BY

WILLIAM FAWCETT, B.Sc., F.L.S.,

Director of Public Gardens and Plantations, Jamaica.

JAMAICA:

GOVERNMENT PRINTING ESTABLISHMENT, 79 DUKE STREET, KINGSTON.

1891

PREFACE.

This Index of Products has been prepared as an indication of the various uses to which plants growing in Jamaica may be put. It has been written somewhat hurriedly in order to have it for reference during the Exhibition.

The arrangement is in alphabetical order of the scientific names with cross references from the common names. After the botanical name comes the common name; then its habitat, followed by a very short description of the plant.

The parts of the plants which are of economic use are arranged in the following order,—root, stem, leaves, flowers, fruit.

Reference has been made, when necessary, to the Bulletin of the Botanical Department. Several works have been consulted, of which the chief are Watt's Dictionary of Economic Products of India, Vols. I. and II., A to C; Bentley and Trimen's Medicinal Plants; Macfadyen's Flora of Jamaica, Vols. I. and II; Kew Bulletin. The last mentioned publication should be in the hands of all who are in any way interested in tropical agriculture.

W. F.

ABRUS PRECATORIUS, Linn.

WILD LIQUORICE, CRAB'S EYES.

Native of India. Naturalised in West Indies.

A twining plant, belonging to the Pea Family *(Leguminosæ)*, with rose-coloured flowers and scarlet seed with black eye.

ROOT has been used as a substitute for liquorice.

LEAVES.—Fresh leaves, steeped in warm castor oil, relieves local pain.

SEEDS.—" Used internally in affections of the nervous system and externally in skin diseases, ulcers, affections of the hair." (Dutt.) One to 3 grains boiled with milk is a tonic; unboiled, purgative and emetic. Decorticated and finely ground, cause purulent ophthalmia, due to a substance *abrin* which is poisonous to the blood, but not in the stomach.

Used by goldsmiths in India as weights; the Arab name *girat* is said to be the origin of the weight *carat*.

Used ornamentally for necklaces, etc. The specific word *precatorius* points to their use for rosaries.

ACACIA CATECHU, Willd.

CATECHU, CUTCH.

Native of India and Burma.

A tree; leaves compound; flowers small, clustered in spikes, white or pale yellow; pod flat, 5 or 6 inches long. *(Leguminosæ.)*

This tree yields a gum, which is a good substitute for gum arabic.

" Catechu" is the resinous extract, obtained by boiling down a decoction of chips of the heartwood.

A solution of Catechu is, by the action of lime or alum, changed into a dull red colour, which constitutes a fairly good dye.

It is not a good tan, as it colours the skin.

Medicinally, Catechu is a powerful astringent. "It may be employed to restrain immoderate discharges in all cases unattended with inflammatory action.

It is given in diarrhœa and leucorrhœa, and combined with the balsam of copaiba, in the chronic stage of gonorrhœa. It forms an excellent tooth-powder." (Macfadyen.)

Cutch consists of the crystals deposited on twigs placed in a boiling decoction from the chopped wood. It is chewed by the Hindoos in combination with betel-nut, betel leaf, and lime.

WOOD: "sapwood yellowish white; heartwood, dark or light red, extremely hard. The wood seasons well, takes a fine polish, and is extremely durable. It is not attacked by white ants or toredo. It s used in India for oil and sugar cane crushers. agricultural implements, bows, spear and sword handles, and wheelwrights' work. A cubic foot weighs about 70lbs." (Watt.)

ACHRAS SAPOTA, Linn.

NASEBERRY, SAPODILLA.

West Indies. Tree with dark green shining leaves and brown fruit. (*Sapotaceæ.*)

WOOD durable.

FRUIT, when ripe and fresh, is sweet, and of a good flavour.

ACROCOMIA SCLEROCARPA, Mart.

GROO-GROO PALM.

A native of the W. Indies, and tropical S. America. A palm, 30 to 45 feet high, prickly stem; leaves pinnate; spathe covered with black prickles; fruit yellow, fleshy, 1½ inch diem. (*Palmæ.*)

Yields a fibre "of remarkable softness and fineness." (Cross.) See Bulletin No. 17.

FRUIT.—Oil used for external application to ease pain.

ADENANTHERA PAVONINA, Linn.

CIRCASSIAN SEEDS.

Native of India. A tree belonging to the Pea Family (*Leguminosæ*) with compound leaves, and bright scarlet seeds.

LEAVES. A decoction is used for chronic rheumatism and gout; as an astringent and tonic in atonic diarrhœa and dysentery.

WOOD. The red heart-wood is used in India as a substitute for red sandal wood. It is hard, closer grained, durable and strong, and is used for house-building and cabinet-making.

Powdered and mixed with water, it relieves prickly heat, and headaches.

SEEDS. Powdered, and rubbed with water, applied for boils, prickly heat, also for headaches.

Powdered and beaten up with borax, forms a good cement.

As ornaments for necklaces, &c., and also as weights.

ADRUE. See CYPERUS ARTICULATUS.

AGAVE MORRISII, Bak.

CORATOE, MAY POLE.

Native of Jamaica. An agave with green leaves, with prickles, and orange-coloured flowers. (*Amaryllideæ.*)

LEAVES yield a fibre, which is not of much value, as it is slightly curled. The juice is diuretic; forms a soapy lather with salt water. "Leaf roasted—cataplasms—maturating . . . Juice, boiled to a thick consistence, spread on leather, as a plaster, to be used in gout." (Dancer.)

POLE—section, used as razor strop.

AGAVE RIGIDA, Mill., var. SISALANA, Perr.

SISAL HEMP.

Native of Central America, naturalized in Florida.

An agave with glaucous leaves, with or without prickles (*Amaryllideæ.*)

Dry, rocky soil suits this plant best. Information on Sisal Hemp will be found in Bulletin No. 15.

Retting is not required for the agaves. Sisal Hemp resists the action of wet, and is therefore useful for cables, rigging-cordage, &c.

AKEE. See CUPANIA EDULIS.

ALBIZZIA LEBBEK, Benth.

SIRIS TREE, WOMAN'S TONGUE.

Native of tropics in old world. Naturalised. A tree with compound deciduous leaves, and long light-coloured pods. It belongs to the Pea-Family (*Leguminosæ.*)

ROOT.—Bark strengthens gums when spongy and ulcerated.

BARK applied to injuries to the eye, and used in tanning. The gum is used to adulterate gum-arabic in calico printing, and in the preparation of gold and silver leaf cloths.

WOOD: weight 40 to 60lbs. per cubic foot. It seasons, works, and polishes well, and it is fairly durable. (Watt). Used for furniture, boats, sugar-cane crushers, oil mills, picture frames, etc.

LEAVES said to be useful in ophthalmia. In India, given to camels as fodder.

FLOWERS used in India as a cooling medicine, and externally for boils, eruptions, etc.

SEEDS are astringent, and the oil extracted from them is thought useful in leprosy. The powdered seeds have been administered in scrofulous enlargement of glands.

ALEURITES MOLUCCANA, Willd.

INDIAN WALNUT, CANDLE NUT.

Tropics. A tree, with simple, often 3-lobed leaves. (*Euphorbiaceæ.*)

Root affords a brown dye.

BARK yields a gum, also found on fruit.

SEEDS. Kernels taste like walnuts.

Nuts strung on strip of bamboo burn like a candle.

A fixed oil is obtained either by boiling bruised seeds or by expression, superior to linseed oil, as a drying oil for paint, and for other purposes connected with the arts.

Used as mordant for vegetable dyes.

Cake used after expression, as fodder for cattle, and also as manures.

Acts as a mild purgative, its action being unattended with either nausea, colic, or other ill effects. It approaches castor oil, and has been found quite as certain in its action, with the advantage of possessing a nutty flavour; dose ½ to 1 oz. (Pharm. of India.)

ALLAMANDA CATHARTICA, Linn.

ALLAMANDA.

Tropical America. A climber, with opposite leaves and large yellow flowers. *(Apocynaceæ.)*

LEAVES considered a valuable cathartic in moderate doses.

ALLIGATOR APPLE. See ANONA PALUSTRIS.

ALLIGATOR WOOD. See GUAREA SWARTZII.

ALLIUM CEPA, Linn.

ONION: Wild in Turkestan. Cultivated everywhere. *(Liliaceæ.)*

Bulb used as food, and medicinally.

ALLSPICE. See PIMENTA OFFICINALIS.

ALOE VERA, Linn.

ALOES.

Northern Africa. Cultivated in tropics. A plant with thick succulent leaves and yellow flowers. *(Liliaceæ.)*

ROOT supposed to be efficacious in colic.

LEAVES contain a bitter juice, which drains from the cut end, and dries slowly in the sun.

JUICE fresh is said to be "cathartic, cooling and useful in fevers, spleen, and liver disease, enlarged lymphatic glands, and as an external applicant in certain eye diseases." (Watt.) Pulp useful for boils, as an emmenagogue, and in veterinary medicine.

JUICE inspissated is the Aloes of commerce. The best form of the drug is Socotra Aloes derived from Aloe Perryi and others. In small doses, stomachic and tonic; in larger doses, purgative, and useful in atonic dyspepsia, jaundice, etc.

The leaves yield a fibre.

AMYRIS BALSAMIFERA, Linn.

MOUNTAIN TORCH WOOD, ROSE WOOD.

Native of Jamaica, Cuba, Venezuela, and N. Granada. A small tree, 6 to 15 feet high, pinnate leaves, and white flowers. The branches, when bruised, emit a strong smell. *(Burseraceæ.)*

WOOD.—Capital posts for going in the earth; they readily split up into strips and are used by the peasantry for torches. (Harrison.)

AMYRIS, sp.

ROSE WOOD

"A hard close-grained wood with aromatic smell, grows to two or three feet in diameter, might be used in the manufacture of small articles, like the far famed sandal wood." (Harrison.)

ANACARDIUM OCCIDENTALE, Linn.

CASHEW.

West Indies and South America. A tree, 30 to 40 feet, with simple leaves, and small flowers. The fruit consists of a nut on the apex of a pear-shaped body formed of the enlarged top of the stalk.

WOOD, red, moderately hard, close grained. Weight 38 lbs. per cubic foot.

BARK may be used for tanning.

JUICE from bark, astringent, used as a flux for soldering metals, and as an indelible marking-ink.

GUM from bark, only slightly soluble in water, obnoxious to insects.

KERNELS are commonly roasted, which improves their flavour. They yield by expression a light-yellow, bland, nutritious oil, superior to olive oil.

The shell of the nut yields by maceration in spirit an oil, called Cardole, which is black, acrid and vesicating,—used as an anæsthetic in leprosy, as a blister in warts, corns, and ulcers, as a local stimulant in psoriasis.

The ripe fleshy stalk is used as a fruit.

ANAGALLIS ARVENSIS, Linn.

POOR MAN'S WEATHER GLASS.

Mountains in tropics. A small herb, with opposite, simple leaves, and blue flowers. *(Primulaceæ.)*

The plant is used in cerebral affections, leprosy, hydrophobia, dropsy, epilepsy, and mania. (Watt.)

ANAMIRTA COCCULUS, W. and A.

COCCULUS INDICUS OF PHARMACY.

India and Malay Is. A climbing shrub. *(Menispermaceæ.)*

SEEDS yield by expression an oily substance, used in the form of an ointment as an insecticide to destroy pediculi, &c., and in some obstinate forms of chronic skin diseases.

They are intensely bitter, which causes them to be sometimes used as a substitute for hops in the manufacture of beer; this bitter principle is poisonous.

ANANAS SATIVA, Linn.

PINE APPLE.

Tropical America. An almost stemless plant with spiny leaves. The flowers are arranged many together into a dense head, the whole developing into a single fruit. *(Bromeliaceæ.)*

LEAVES yield fibre, strong yet fine.

Fresh juice is an anthelmintic.

FRUIT used fresh, stewed, or preserved.

Juice is said to "allay gastric irritability in fever." (Watt.) It is diuretic, diaphoretic, and refrigerant, antiscorbutic, and useful in jaundice.

ANDIRA INERMIS, Kth.

BASTARD CABBAGE BARK TREE.

"Very durable as posts. The trunk is generally straight, and reaches a diameter of 12 inches; the smaller ones are much used for rafters." (Harrison)

"Plentiful up the valleys, notably those of the Rio Grande and Plantain Garden River. It is a small straight growing tree, wood used as posts, and, where large enough to be sawn, as roofing timber." (Hooper.)

ANDROPOGON CITRATUS, D. C.

FEVER GRASS. LEMON GRASS.

India and Malay Islands. A coarse grass with lemon-scented leaves. *(Gramineæ.)*

LEAVES yield lemon grass oil, value at 1s. 4d. per ounce. It is used in manufacture of Eau de Cologne, and soaps

In the Indian Pharmacopœia, the oil is officinal, and is considered "stimulant, carminative, anti-spasmodic, and diaphoretic; locally applied it is a rubefacient. It is recommended to be administered in flatulent and spasmodic affections of the bowels and in gastric irritability. In cholera it has been spoken of as a remedy of great value." Infusion of the leaves is also used. (Watt.)

ANDROPOGON MURICATUS, Retz.

KHUS-KHUS GRASS.

India. A grass with aromatic-scented leaves and roots. *(Gramineæ.)*

ROOTS contain a resinous substance with an odour like myrrh, and a volatile oil, used as a perfume.

They are made into mats, fans, baskets.

An infusion is given as a gentle stimulant diaphoretic.

ANNATTO. See BIXA ORELLANA.

ANONA CHERIMOLIA, Mill.

CHERIMOYA.

Ecuador to Peru, and hills in Jamaica. A tree with inconspicuous flowers, and large, smooth, green fruit. *(Anonaceæ.)*

FLOWERS "are put into snuff, as a substitute for the Tonquin bean, for the purpose of giving a grateful flavour." (Macfadyen.)

FRUIT, being somewhat acid, is very agreeable.

ANONA MURICATA, Linn.

SOUR SOP.

West Indies. A small tree with large, green fruit covered with soft prickles. *(Anonaceæ.)*

ROOT—decoction, antidote against fish-poison. (Dancer.)

FRUIT—diuretic.

ANONA RETICULATA, Linn.

CUSTARD APPLE.

Tropical America. A small tree. *(Anonaceæ.)*

WOOD weighs 40lbs. per cubic foot.

BARK is an astringent, and tonic; of young twigs, yields fibre.

LEAVES and young twigs used for tanning. The leaves also yield a kind of indigo.

FRUIT unripe yields a black dye Ripe, it is said to be anti-dysenteric and vermifuge. As a fruit, it is not much relished, being too luscious.

ANONA SQUAMOSA, Linn.

SWEET SOP.

Tropical America A small tree. *(Anonaceæ.)*

ROOT a drastic purgative, administered in acute dysentery, spinal diseases.

LEAVES, immature fruits, and seeds contain a principle fatal to insects. The leaves are often rubbed on floors, etc., in houses to get rid of insects.

Leaves are applied for extraction of guinea worm, to unhealthy ulcers—an anthelmintic.

FRUIT, unripe, dry, powdered, and mixed with flour, used to destroy vermin.

Ripe fruit, agreeable, good for digestion.

WOOD is soft, close-grained. Weight 46lbs per cubic foot.

ANONA PALUSTRIS, Linn,

ALLIGATOR APPLE, CORK WOOD.

W. Indies and tropical S. America. A small tree growing in marshes. *(Anonoceæ.)*

WOOD is very light, used as floats for fishing nets, and as stoppers for mouths of vessels of calabash.

ARACHIS HYPOGÆA, Linn.

PINDAR, EARTH NUT, GROUND NUT.

Brazil. A prostrate annual herb, belonging to the Pea Family *(Leguminosæ)*, with pinnate leaves and yellow flowers. When the pod begins to form, the stalk curves over and buries the pod in the ground where it ripens. More than 100,000 acres are devoted to the cultivation of this plant in India and immense tracts in W. Africa. It requires a dry, sandy soil.

SEEDS afford on expression an oil, which resembles olive-oil, and is used as a substitute for it, both medicinally and for alimentary purposes.

After the expression of oil, the residue may be made into meal which is richer than peas, and even lintels, in flesh-forming constituents, and contains more fat and phosphoric acid. (Muter.) The cake is also recommended for cattle feeding.

In the United States, the nuts are pounded up in a mortar, and are said to make an agreeable chocolate.

Roasted in the shell, the nuts can be used at dessert.

LEAVES and branches are excellent fodder, and the hay increases the milk of cows.

ARECA CATECHU, Linn.

ARECA NUT, BETEL PALM.

Tropical Asia. An elegant palm, with slender stems, attaining a height of 80 feet, with a crown of pinnate leaves.

NUTS, young,—astringent, useful in diarrhœa and urinary disorders. The dried nuts, chewed, produces stimulant and exhilarting effects on the system. Powdered seeds are anthelmintic for dogs.

SPATHE may be used for paper-making; also for bags, caps, &c.

WOOD used for furniture, &c. Weight 57 lbs per cubic foot.

ARGEMONE MEXICANA, Linn.

MEXICAN POPPY.

Tropical America and W. Indies. A prickly annual herb, with yellow juice, and yellow flowers, belonging to the Poppy Family. *(Papaveraceæ.)*

JUICE said to be useful for dropsy, jaundice, and cutaneous affections; diuretic; healing to ulcers.

SEEDS have narcotic properties. They yield on expression an oil, used as an aperient anodyne, and hypnotic. "In stomach complaints, the usual dose is 30 drops on a lump of sugar." (Watt.)

ARISTOLOCHIA ODORATISSIMA, Linn.

CONTRAYERVA.

Native of Jamaica, Central America, and Venezuela. A twining plant, leaves heart-shaped, 4 or 5 inches long; flower curved, tubular, inflated, with an expanded and broad lip. *(Aristolochiaceæ.)*

ROOT—Infusion or decoction, diuretic, purgative, stomachic, emmenagogue. Used also as a medicine for horses.

ARRACACIA XANTHORHIZA, Bancr.

ARRACACHA.

Mountains of Tropical America. Cultivated at Hill Garden, Cinchona, Jamaica.

A perennial herb, belonging to the Carrot Family *(Umbelliferæ)*, with tuberous root-stocks.

TUBERS are used for food. The flavour has been compared to a combination of parsnip and potato.

ARROWROOT. See MARANTA ARUNDINACEA.

ARROWROOT, SPANISH. See CANNA EDULIS.

ARTOCARPUS INCISA, Linn fil.

BREAD FRUIT.

East Indies and Polynesia. A tree with milky sap, and large fruit. *(Urticaceæ.)*

ARTOCARPUS INTEGRIFOLIA, Linn. fil.

JACK FRUIT.

East Indies and Polynesia. A tree, with milky sap, and large fruit. *(Urticaceæ.)*

BARK yields gum, used as a cement and bird-lime; and also a fibre.

JUICE applied externally to glandular swellings to promote suppuration.

WOOD yields on boiling a yellow dye Timber used for carpentry, cabinet-work, etc. "Yellow, hard, takes an excellent polish, is beautifully marked, and is one of the handsomest furniture woods." (Warden.) Weight 40lbs. per cubic foot.

ASCLEPIAS CURASSAVICA, Linn.

RED HEAD, WILD IPECACUANHA.

Native of W. Indies, Central and tropical S. America. An erect, perennial herb with orange and crimson-coloured flowers, and large seed vessels; seeds with long white silky hair. *(Asclepiadeæ.)*

ROOT possesses emetic and cathartic properties; it is purgative, and subsequently astringent; remedy in piles and gonorrhœa.

LEAVES—juiçe anthelmintic; useful in arresting hœmorrhages, and in obstinate gonorrhœa; sudorific.

FLOWERS—juice, a styptic.

AVERRHOA BILIMBI, Linn.

BILIMBI TREE.

Cultivated in E. Indies. A small tree, with reddish purple flowers, cylindrical fruit with 5 rounded lobes. *(Geraniaceæ.)*

FLOWERS made into preserves.

FRUIT used in curry, and preserved in sugar.

Juice, made into a syrup as cooling drink in fevers; and used externally in cutaneous diseases.

Juice also used to take iron moulds from clothes, and ink and other stains from furniture.

AVERRHOA CARAMBOLA, Linn.

CARAMBOLA, CARAMBA.

Cultivated in E. Indies. A small tree, with yellowish-purple flowers, fruit acutely 5-angled. *(Geraniaceæ.)*

ROOT, *leaves, and fruit*, used as cooling medicine.

FRUIT—unripe, astringent, and used as an acid in dyeing, probably acting as a mordant.

Ripe—antiscorbutic.

Juice removes iron-moulds from linen. Made into curries, pickles, and preserves.

WOOD light-red, hard, close-grained. Weight 40lbs. per cubic foot.

BAMBUSA VULGARIS, Wendl.

Bamboo.

Native of East Indies. A gigantic grass, with woody stems, 20 to 50 feet high. *(Gramineæ.)*

The fibre of the Bamboo is an excellent paper material. In China, it is the principal, if not the only, material for paper-making. The Chinese use the native bamboo, which they split into lengths of 3 or 4 feet, and place in a layer in a tank. This is covered with lime, and alternate layers of bamboo and lime, are so placed, until the tank is full. Water is run in to cover the whole, and left for 3 or 4 months, when the bamboo has become rotten. The soft bamboo is pounded in a mortar into a pulp, mixed with water, and then poured on square, sieve-like moulds. The sheets are allowed to dry on the mould, then placed against a hot wall, and finally exposed to the sun. Mr. Routledge advocated the use of young shoots, but one difficulty is that cutting them weakens the stock; in fact, if all the young shoots are cut for three successive years, the stock dies. At Lacovia, bamboo is crushed, and exported in short lengths as packing for cylinders.

The young shoots, freed from the sheaths, are used in India in curries, pickles, and preserves. The very young shoots are not unlike asparagus.

"It would occupy a volume even to enumerate by name all the uses to which the mature bamboo stems are put. Suffice it to say that to the inhabitants of the regions where the bamboo luxuriates, it affords all the materials required for the erection and furnishing of the ordinary dwelling-house." (Watt.)

Mr. A. R. Wallace, in "Tropical Nature," details some of "the endless purposes to which the bamboo is applied in the countries of which it is a native, its chief characteristic being that in a few minutes it can be put to uses which, if ordinary wood were used, would require hours or even days of labour. There is also a regularity and a finish about it which is found in hardly any other woody plant, and its smooth and symmetrically ringed surface gives an appearance of fitness and beauty to its varied applications."

BANANA. See Musa paradisiaca, var. sapientum.

BARBADOS PRIDE. See Cæsalpinia pulcherrima.

BAUHINIA VARIEGATA, Linn.

Native of India and Burma. Naturalised in Jamaica. A small tree, leaves 2-lobed; 4 inches diam.; flowers rosy-white, one petal with a purple blotch at base, $1\frac{1}{2}$ to 2 inches; pod flat, 3 to 5 inches long. *(Leguminosæ.)*

Root—decoction given in dyspepsia and flatulency.

Bark, used in dyeing and tanning. It is, medicinally, alterative, tonic, and astringent; a vermifuge; useful in scrofula, leprosy, skin diseases, and ulcers.

GUM—brown-coloured, a small proportion soluble in water.

WOOD, hard and serviceable, but small size.

BUDS, "dried, used as a remedy for piles and dysentery. They are considered by the natives (of India) as cool and astringent, and are useful in diarrhœa and worms." (Baden Powell.)

"FLOWERS given with sugar, as a gentle laxative; and the bark, flowers, or root, triturated in rice-water, as a cataplasm to promote suppuration." (Watt.)

BAYBERRY. See PIMENTA ACRIS

BEAN. See PHASEOLUS.

BETEL PALM. See ARECA CATECHU.

BILIMBI TREE. See AVERRHOA BILIMBI.

BIRCH, WEST INDIAN. See BURSERA GUMMIFERA.

BISSY. See COLA ACUMINATA.

BITTER ASH. See PICRÆNA EXCELSA.

BITTER BUSH. See EUPATORIUM NERVOSUM.

BITTER DAN. See SIMARUBA GLAUCA.

BITTER WOOD. See PICRÆNA EXCELSA, & SIMARUBA GLAUCA.

BIXA ORELLANA, Linn.

ANNATTO.

Native of W Indies and Tropical America. A low tree, 10 feet high, with large rosy-coloured flowers; seed-vessels spiny, and seeds covered with a coloured pulp. *(Bixineæ.)*

SEEDS are exported from Jamaica in large quantities, and the colouring matter (annatto) removed in England. In Cayenne, and Guadeloupe, the annatto is made up into cakes for export. Mr. J. J. Bowrey, Island Chemist in Jamaica, has invented a method of obtaining a superior kind of Annatto in powder. See Bulletin VII., 4. In Europe, this colouring matter is used for cheese, butter, soaps, &c.

Seeds in medicine, are cordial, astringent, and febrifuge.

BARK yield a fibre.

WOOD soft; the friction of two pieces is used to produce fire.

BLACKBERRY. See RUBUS ALPINUS and R. JAMAICENSIS.

BLOOD WOOD. See LAPLACEA HÆMATOXYLON.

BLUE GUM. See EUCALYPTUS GLOBULUS..

BOCAGEA LAURIFOLIA, B. & H.

WHITE LANCEWOOD.

Native of Jamaica, Cuba, Porto Rico. A tree of moderate height with small white flowers, and leaves pointed at both ends. *(Anonaceæ.)*

Stem, straight, light, and tough. "Lancewood Spars" are exported for use by coachbuilders. Probably it would prove remunerative to pay some attention to planting and encouraging the growth of lancewood.

BOCAGEA VIRGATA, B. & H.

Black Lancewood.

Native of Jamaica, Cuba, Haiti. Similar to White Lancewood, but the leaves are ovate; used for the same purposes.

"Grows straight to 20 or 30 feet in height, and 8 to 11 inches at the butt, possesses great elasticity; much used in carriage building, for which purpose it is exported." (Harrison.)

"On the coast ranges, especially on the south side. A tall slim tree, with a diameter of eight inches. The timber is very elastic, and on that account it is exported for conversion into carriage shafts; Jamaica spars fetching higher prices in the home market than similar produce from other sources. Exports past 20 years, 204,000, valued at £31,275." (Hooper.)

BOCCONIA FRUTESCENS, Linn.

Celandine, Parrot Weed, John Crow Bush.

Native of W. Indies and tropical America. An erect shrubby plant, with large oak-like leaves, and bunches of petal-less flowers. *(Papaveraceæ.)*

Root "scraped and beat up into a pulp, is an excellent application to foul ulcers. The juice of the root has been employed in cases of chronic ophthalmia, to remove warts and fungous flesh, as an application for tetters and ring-worm." (Macfadyen.)

Leaves—the juice is used for ophthalmia. They are also used for rubbing on floors of houses, as they get rid of insects.

BŒHMERIA NIVEA, Hook & Arn.

China Grass, Ramie.

Native of Southern Asia.

The plant belongs to the nettle tribe *(Urticaceæ)*, and grows best in rich, fertile soil, with plenty of water.

The fibre, obtained from the young shoots, is one of the strongest and most beautiful. "It is glossy, tough and lasting, combining to some extent the appearance of silk with the strength of flax." (Mueller.) No machine or process has yet been devised by which the fibre may be extracted easily and cheaply. The fibre is contained in the bark which surround a hard woody core. It is easy enough to strip off the bark in "ribbons," but a resinous substance becomes hard, and complicates the process of extraction. "In 1871 a reward of £5,000 was offered by the Indian Government for a good extracting machine for this fibre; but although several competitors came forward, the prize was awarded to no one." (Watt.) There were trial competitions carried on at the Paris Exhibition, but with no satisfactory result.

BOTTLE GOURD. See LAGENARIA VULGARIS.

BOX WOOD. See VITEX UMBROSA.

BREAD FRUIT. See ARTOCARPUS INCISA.

BREAD NUT. See BROSIMUM ALICASTRUM.

BRAZILETTO. See PELTOPHORUM LINNÆI.

BROAD LEAF. See TERMINALIA LATIFOLIA.

BROMELIA PINGUIN, Linn.

PINGUIN.

Native of W. Indies and tropical America. A plant belonging to the same family as the Pine-Apple (*Bromeliaceæ*), but the fruits remain separate on the stalk.

LEAVES yield a fibre at a percentage of 2·7, which is too small to be profitable. (Morris.) See Bulletin XVII. 10.

BROOM WEED. See SIDA CARPINIFOLIA, and S. RHOMBIFOLIA.

BROSIMUM ALICASTRUM, Sw.

BREADNUT.

Native of Jamaica and Central America. A high tree; leaves simple, 3 –6 inches long; flowers minute, crowded on receptacles. (*Urticaceæ.*)

WOOD.—"This is an excellent timber tree and grows abundantly in the interior, generally straight, with a diameter of about 18 inches; it makes capital boards, takes a high polish and makes a beautiful flooring.

NUTS.—The tree bears abundant nuts, which are readily eaten by stock of all kinds. Horses and cattle are also very fond of the leaves." —(*Harrison.*)

"On the interior slopes of the northern coast range, and notably on the levels of the hills in St. Catherine, Clarendon and Manchester above the railway line. A tall erect tree, up to 80 feet, with a diameter of two feet. The heart-wood, especially in the roots, has a rich brown colour, which, with its very durable qualities, makes it prized for floorings and ornamental work of all kinds. It yields a bountiful supply of nuts, which form a valuable fodder, as also do the leaves." (*Hooper.*)

BRYA EBENUS, DC.

WEST INDIAN EBONY, COCCUS WOOD.

Native of Jamaica and Cuba. A small tree 15–20 feet high, with drooping branches, small leaves, clustered yellow flowers and two-jointed pod. (*Leguminosæ.*)

"Found in quantity at the base of the Clarendon Hills and elsewhere near the south coast. A small tree with rugged grey bark, rarely found with diameter of over eight inches. Has a hard deep-coloured heartwood, close-grained, and on this account has been exported. Generally known in commerce as Coccus wood." (*Hooper.*)

BRYOPHYLLUM CALYCINUM, Salisb.

LEAF OF LIFE.

Native of Asia. A succulent herb, leaves producing new plants from the indentations; flowers reddish yellow. (*Crassulaceæ.*)

LEAVES applied to contused wounds, boils, ulcers, &c.

BUCIDA BUCERAS, LINN.

WILD OLIVE, OLIVE BARK TREE.

Native of W. Indies and Panama. A tree, 20 to 30 feet high; leaves rounded, at the end of branches; flowers without petals; berry small, ¼ inch, crowned with the persistent calyx. (*Combretaceæ.*)

WOOD: used by cabinet-makers.

BARK: used "in tanning of sole-leather." (Sloane.)

BUCIDA CAPITATA, Vahl.

YELLOW SANDERS, WILD OLIVE, NEGRESSE.

Native of Jamaica, Cuba, Guiana, and Brazil. A tree, 30 to 60 feet high; leaves roundish, crowded at ends of twigs: flowers without petals; fruit of size and shape of olive.

WOOD: "chiefly found to perfection in the interior where it grows straight and 3 or 4 feet in diameter. It saws freely and makes a beautiful board, taking a high polish. The wood is of a light yellow colour with satin graining, and is highly prized in cabinet work, where it sets off dark woods." (Harrison.)

"Chiefly in the interior of the central districts. A large tree with diameter up to three feet, largely used in cabinet work on account of its light colour and satiny grain." (*Hooper.*)

BARK "has an aromatic astringent and bitter taste, and a decoction of it, we are informed, has been employed as a remedy for the complicated diseases, resembling constitutional syphilis, to which the African race is subject." (Macfadyen.)

BUCIDA SP.

GREY MOUNTAIN SANDERS.

Native of Jamaica.

WOOD.—"This is a very good timber, saws readily, makes a fine board, darker in colour than the yellow sanders and not so satin-like in appearance, but takes fine polish, grows to about three feet in diameter." (*Harrison.*)

BULLET TREE. See DIPHOLIS MONTANA.

BULLET TREE, MAHOGANY. See SAPOTA SIDEROXYLON

BULLY TREE. See DIPHOLIS MONTANA.

BURSERA GUMMIFERA, Linn.

WEST INDIAN BIRCH.

Native of W Indies, Bahamas, and tropical America. A high tree; eaves pinnate, flowers very small, appearing before the leaves.

Wood.—"A tree of the coast and the coast ranges. A smooth barked, erect, deciduous tree, from its clean stem and its habit of branching only at the top, adapted for live telegraph posts. As a timber used in coopering." (Hooper.)

All parts of the tree produce a gum, capable of being substituted for gum-mastic as a transparent varnish. It might be given in form of pills, as a substitute for copaiba, in diseased discharges from the mucous membranes. (Macfadyen.)

BUTTER WEED. See Erigeron canadense.

BUTTON WOOD. See Conocarpus erecta.

BYRSONIMA CORIACEA, Dc.

Lotus-Berry Tree.

Native of W. Indies. A tree 20 to 30 feet high: leaves simple, flowers golden-yellow: fruit, yellow, size of a small cherry.

Fruit edible.

BYSSY. See Cola acuminata.

CABBAGE BARK TREE, BASTARD. See Andira inermis.

CACAO. See Theobroma Cacao.

CÆSALPINIA BONDUCELLA, Fleming.

Grey Nicker Seed, Bonduc.

Cosmopolitan in the tropics. A woody climber, belonging to the Pea Family *(Leguminosæ)*; leaves twice-pinnate with hooked prickles by which the plant climbs: flowers yellow; pod dry, orange-brown covered with spines; seeds lead-coloured.

Root—bark, and also the

Seeds considered tonic, antipyretic and antiperiodic.

Seeds used for necklaces, etc.

CÆSALPINIA BONDUC, Roxb.

Yellow Nicker Seeds.

West Indies, East Indies, Polynesia, Malay Isles. A plant like the preceding, but of larger size, less hairy, and with yellow seeds.

Seeds used for necklaces, etc.

CÆSALPINIA CORIARIA, Willd.

Divi-divi.

Native of W. Indies and tropical America. A small crooked tree; leaves twice-pinnate; flowers fragrant, white; pod flat, incurved. This tree is worth cultivating, it grows in hottest and driest places.

Wood of little value.

Pods rich in tannin. Exported to a very small extent. If seeds not removed, the oil they contain induces fermentation; this might be obviated, and charges for freight lessened by reducing the pods to a powder, or preparing an extract from fresh pods. England imports about 4,000 tons a year, and 12,000 tons of sumach, but the latter is being replaced by the cheaper divi-divi. (*Watt.*)

Von Mueller recommends it for cultivation in salt-marshes.

"Powder of the pods astringent, anti-periodic, tonic. Dose one to two drachms as an antiperiodic." (*Ward in Watt.*)

CÆSALPINIA PULCHERRIMA, *Sw.*

BARBADOS PRIDE.

Tropics. A prickly shrub, 5 to 10 feet high, leaves compound; flowers very showy, red or yellow, with very long stamens. (*Leguminosæ.*)

WOOD, charred, yields an ink.

LEAVES AND FLOWERS: Infusion—a powerful emmenagogue.

LEAVES—A purgative, used as a substitute for senna.

SEEDS, powdered also medicinal.

CAJANUS INDICUS, *Spreng.*

NO EYE PEA (small form), CONGO PEA (large form).

Tropics. A shrub 6 to 12 feet high; leaves compound, with 3 leaflets; flowers yellow; pod, compressed, constricted between seeds (*Leguminosæ.*)

LEAVES, tender, chewed in cases of apthæ and spongy gums. Mixed in a paste with the pulse, and applied warm, checks the secretion of milk.

SEEDS, esteemed as food, but apt to prove irritant and laxative, properties which may be minimised by freeing from husk. (*Church.*)

CALABASH. See CRESCENTIA CUJETE.

CALOPHYLLUM CALABA, Jacq.

SANTA-MARIA.

Native of W. Indies and tropical America. A lofty tree; leaves simple, feather-veined, with delicate veins; flowers small, white, fragrant; berry 1 inch diameter. (*Guttiferæ.*)

WOOD—"This wood is very abundant and on that account much used for building, although it is not considered a durable wood; shingles of an inferior class are split from this wood. The growth in humid localities is about 4 feet in diameter at the butt. I have seen trees 150 feet high and as straight as a ship's mast." (*Harrison.*)

"Especially in the south of Trelawny, but generally at an altitude of from 2,000 to 3,000 feet. Tall straight growth up to 100 feet and more, with a diameter of three feet. Yields a second-class timber and splits into shingles inferior to several other kinds. It is an important forest tree" (*Hooper.*)

CALOTROPIS PROCERA, R. Br.

French Cotton.

Native of tropical Africa, N. India, and Persia. Naturalised in West Indies and Central America.

A shrub, 5 to 15 feet high; leaves large, opposite, pale glaucous green; flowers pink, shaded and dotted with purple; pod-like fruit: seeds numerous, each with a tuft of white silken hairs at one end.

Bark of root, known in India as "Mudar,"—alterative, tonic, diaphoretic, and in large doses emetic. Beneficial in obstinate cutaneous diseases, syphilitic affections, dysentery, diarrhœa, and chronic rheumatism.

Stem yields fibre. Sap yields a kind of gutta-percha, but unfortunately it is a good conductor of electricity, and therefore unsuited for manufacture of cables. Bark yields a rich white bast fibre.

Seeds—Hairs used in fancy work.

CAMELLIA THEIFERA, Griff.

Tea.

Native of Assam. A shrub, with large, dark-green leaves, and white flowers. (*Ternstrœmiaceæ.*)

Leaves are picked when opening from the bud, and cured, the difference in the size of the leaves and methods of curing, determining the different kinds of tea.

CANDLE NUT. See Aleurites moluccana.

CANDLE WOOD, YELLOW. See Cassia emarginata.

CANELLA ALBA, Murray.

Wild Cinnamon. White Canella.

Native of W. Indies, Bahamas, Florida.

A tree, 30 or 40 feet high, with simple leaves; flowers small, of a pale violet colour, and sweet aromatic smell. (*Canellaceæ.*)

Bark—An aromatic stimulant and slight tonic. Useful in dyspepsia, gout, rheumatism, syphilis, scurvy.

Used also as a condiment.

The principal constituent is a volatile oil.

CANELLA, RED. See Cinnamodendron corticatum.

CANELLA, WHITE. See Canella alba.

CANNA EDULIS, Ker.

Spanish Arrow Root. Tous les Mois.

Native of tropical S. America and Trinidad.

An herbaceous perennial, with a creeping root-stock; flowering stems 6 to 8 feet high; leaves very large; flowers red.

ROOT-STOCKS grated into a pulp. The pulp is washed, the liquid strained, and allowed to settle. The water is decanted, and the starch dried. The grains of starch are much larger than any other starch. They look somewhat flattened, and are oblong in form.

Tous les Mois is nutritious and wholesome.

It is a demulcent in urinary and bowel complaints.

CANNABIS SATIVA, Linn.

HEMP.

Native of Asia. The Hemp plant is an annual growing to a height of 4 to 10 feet. It belongs to the Nettle family. (*Urticaceæ.*)

It is possessed of narcotic properties, and in India the dried plant is smoked under the name of Gunjah, and pounded in water to make a drink under the name of Bhang.

A resin exudes from the plant, and is known as Churras. In small quantities it produces excitement, and in increasing and continued doses, delirium, catalepsy and insanity.

The fruit (commonly known as hemp-seed) contains a single oily seed, which yields on compression the well known hemp-oil.

The bark contains the fibre which makes the plant so valuable. Good well drained, dampish soil is required for its cultivation. Russia and Poland produce very large quantities, but the Italian is considered superior. To produce the best fibre the seed is sown close, which prevents branching.

CAPSICUM FRUTESCENS, Linn.

CAYENNE PEPPER.

Native of W. Indies and Tropical America. A shrub, with red or yellow, conical berry, ½ to 1 inch long. (*Solanaceæ.*)

FRUIT: "The great part of the so-called Cayenne Pepper is made from it, but the name is given also to the product of other peppers." (*DeCandolle.*) Used in pickles, curries, &c.

"Used as medicine in typhus and intermittent fevers and in dropsy; they are regarded as stomachic and rubefacient." (Watt.)

CARAMBA AND CARAMBOLA. See AVERRHOA CARAMBOLA.

CARDIOSPERMUM HALICACABUM, Linn.

HEART PEA.

Tropics. A climbing annual plant, with tendrils; leaves compound, twice ternate; flowers small white: seed vessel inflated, globular, 1 inch long. (*Sapindaceæ.*)

ROOT: "An emetic, laxative, stomachic, and rubefacient. It also possesses diaphoretic, diuretic, and tonic properties." (Watt.) "The decoction is mucilaginous and somewhat nauseous, and has the reputation of being lithontriptic." (Macfadyen.)

LEAVES: In the East Indies the young shoots and leaves are cooked as a vegetable They are administered in pulmonic complaints, and mixed with castor oil, are internally employed in rheumatism and lumbago. They are also applied externally.

SEEDS: "Bruised in water and applied externally, they relieve the pain in gout, and are useful in chronic affections of the joints; and mixed with sugar they may be given for cough." (Macfadyen.)

CARICA PAPAYA, Linn.

PAPAW.

Native of W. Indies and Central America. Naturalised throughout tropics. A small tree, with milky juice, generally branchless; leaves large, lobed; flowers small; fruit yellow, size of a small melon. (*Passifloraceæ.*)

FRUIT: Juice of *unripe* fruit possesses authelmintic properties, expelling lumbrici. It is useful in dyspepsia as a vegetable substitute for pepsine. Juice also applied in psoriasis and skin affections of like character; in ring worm. The unripe fruit is cooked as a vegetable and the ripe fruit as dessert has the same effect, and acts as a mild chologogue and purgative; hence its use for piles, enlarged liver and spleen. (*Watt.*) The active principle, *papaine*, has been separated in the form of a dry white powder by Mr. J. J. Bowrey, Island Chemist in Jamaica. Papaine is extensively used in France and Germany. See Bulletin, No. 9.

The digestive property of the juice and of the fresh leaves is made use of to render meat tender, and facilitate the process of cooking.

CARYOPHYLLUS AROMATICUS, Linn.

CLOVE.

Native of Moluccas. Cultivated in Jamaica. A small tree, of which the unopened flower-buds form the cloves of commerce. (*Myrtaceæ.*)

FLOWERS: Buds and stalks abound in an essential oil. Cloves are aromatic, stimulant, and carminative; used in atonic dyspepsia and in gastric irritability.

CARYOTA URENS, Linn.

WINE PALM. KITTUL FIBRE PALM.

Native of East Indies. Cultivated in Botanic Gardens, Jamaica. A beautiful palm with smooth ringed stem. (*Palmæ.*)

LEAVES "give the Kittul Fibre, which is very strong and is made into ropes, brushes, brooms, baskets, and other articles; the fibre from the sheathing leaf-stalk is made into ropes and fishing-lines" (Gamble), and is said to be suitable for paper manufacture.

"This tree is highly valuable to the natives of the countries where it grows in plenty. It yields them, during the hot season, an immense quantity of toddy or palm wine. I have been informed that the best trees will yield at the rate of 100 pints in the 24 hours. The sap in some cases continues to flow for about a month. When fresh, the toddy is a pleasant drink, but it soon ferments, and when distilled becomes arrack, the gin of India. The sugar called jaggery is obtained by boiling the toddy. The pith or farinaceous part of the trunk of old trees is said to be equal to the best sago; the natives make it into bread, and boil it into thick gruel." (Roxburgh.)

WOOD, strong and durable, used for agricultural purposes, water conduits, and buckets.

CASHAW. See PROSOPIS JULIFLORA.

CASHEW. See ANACARDIUM OCCIDENTALE.

CASSAVA. See MANIHOT UTILISSIMA.

CASSIA ALATA, Linn.

RING-WORM SHRUB.

Tropics. A shrub, 6 to 10 feet high; leaves compound, flowers large, showy, yellow; pods 4-winged, 5 inches long. (*Leguminosæ.*)

LEAVES, bruised, and mixed with an equal weight of simple ointment, is described as being almost a specific for ring-worm, but it is still more efficacious, if the bruised leaves are rubbed in with lime-juice. Taken internally, they act as an aperient.

"FLOWERS and young LEAVES beat into a pulp make an excellent poultice for the superficial sores which follow some varieties of impetigo and rupia." (Macfayden.)

CASSIA EMARGINATA, Linn.

YELLOW CANDLE WOOD.

Native of W. Indies.

A shrubby tree, 10-15 feet high; with pinnate leaves, yellow flowers, and straight pod. (*Leguminosæ.*)

WOOD—"This is a dyewood, for which purpose some has been exported. It is hard but does not grow large enough to saw for timber." (Harrison.)

CASSIA FISTULA, Linn.

PURGING CASSIA.

Native of E. Indies, China, Malay Isles.

A handsome tree, with pinnate leaves, large yellow flowers, several together on long pendulous stalks, and pods 1-2 feet long, cylindrical, with many seeds immersed in a dark-coloured pulp. (*Leguminosæ.*)

PODS are exported. The pulp is a mild laxative, and is chiefly used as an ingredient of the preparation—Confection of Senna.

CASSIA OCCIDENTALIS, Linn.

WILD COFFEE.

Tropics. A shrub, 3 or 4 feet high; leaves compound; flowers yellow, ½ inch long; pod nearly cylindrical, 2 to 4 inches long. (*Leguminosæ.*)

ROOT—Diuretic.

LEAVES—Decoction, taken internally and applied externally, in cure of itch and other cutaneous diseases, also of mange.

SEEDS used for ring-worm. Roasted they are a good substitute for coffee.

CASTOR OIL. See RICINUS COMMUNIS.

CASUARINA EQUISETIFOLIA, Forst.

CASUARINA, BEEFWOOD OF AUSTRALIA.

Native of Australia and E. Indies. Cultivated in Jamaica. A large tree with leafless, drooping branchlets. (*Casuarinaceæ.*)

BARK, astringent, useful in chronic diarrhœa and dysentery.

WOOD, hard and heavy. Casuarina seems to coppice well, and is an important tree for fuel. (Gamble.) Valued for steam-engines, ovens, &c.

CATALPA LONGISSIMA, Sims.

YOKE WOOD, MAST WOOD, FRENCH OAK.

Native of Jamaica, Haiti, St. Thomas. A large, handsome tree; leaves simple, opposite; flowers delicate rosy-white; pod pendulous, narrow, 2 feet long. (*Bignoniaceæ.*)

WOOD. One of the most useful and best timbers in the island.

"This wood grows abundantly on the south side; it is sawn up into boards and scantlings; it is very durable and not too hard for general purposes; it grows to about 3 feet in diameter." (Harrison.)

"Along the coast, notably in the south and southeast: rarely found above 1,000 feet elevation. A fine well-grown tree, with diameter up to 4 feet and height of 80 feet. It is easily sawn and is used for general purposes in the form of boards and planks" (Hooper.)

PODS have been used for plaiting hats, &c.

CATECHU. See ACACIA CATECHU.

CAYENNE PEPPER. See CAPSICUM FRUTESCENS.

CEDAR, BASTARD. See GUAZUMA TOMENTOSA.

CEDAR, BERMUDA. See JUNIPERUS BARBADENSIS.

CEDAR, JUNIPER. See JUNIPERUS BARBADENSIS.

CEDAR, WEST INDIAN. See CEDRELA ODORATA.

CEDRELA ODORATA, Linn.

CEDAR, WEST INDIAN CEDAR.

Native of West Indies, and Central America.

A high tree; leaves pinnate: flowers pale-yellow with a peculiar odour; seed-vessel opening by 5 valves from the top; seeds flat, winged. (*Meliaceæ.*)

Wood: "This is a well known wood, it cuts freely, is not hard, very durable, and is perhaps more extensively used for general building purposes, away from the coast, than any other kind of indigenous wood. A great many shingles are made of cedar. In the interior it grows to 4 and even 6 feet in diameter and 70 to 80 feet high." (Harrison).

Bark: gum, resembling gum arabic, obtained by making incisions.

"Generally over the island, especially at altitudes varying from 1,000 to 4,000 feet. Only slightly gregarious. Chiefly dotted over open pastures and along watercourses. Grows up to a large size; six feet in diameter in the open, rarely over half that diameter at high elevations, where, in sheltered narrow vales, it has tall straight stems. It is a light handsome wood, in most general request for furniture, roofing shingles, interior of houses, cigar boxes and ornamental work. Exported to Europe and elsewhere." (Hooper.)

CEIBA, See Eriodendron Anfractuosum.

CELANDINE. See Bocconia frutescens.

CERASEE. See Momordica Balsamina.

CHEW STICK. See Gouania domingensis.

CHERIMOYA. See Anona cherimolia.

CHINA GRASS. See Boehmeria nivea.

CHOCHO. See Sechium edule.

CHOCOLATE TREE. See Theobroma cacao.

CHRYSOBALANUS ICACO, Linn.

Cocoa Plum.

Native of W. Indies, Bahamas, and Tropical America and Africa.

A shrubby tree, 12 feet high; leaves roundish; flowers white; fruit, brownish-purple, size of a plum. (*Rosaceæ.*)

Roots and Leaves: "An astringent bath, recommended in leucorrhœa and blennorhœa, is prepared from the leaves and roots."

Fruit "considered astringent and to be useful in bowel complaints.

Prepared with sugar, it forms a favourite conserve with the Spanish colonists, and large quantities are annually exported from Cuba.

The kernels yield a fixed oil, and an emulsion made with them, is said to be useful in dysentery." (Macfadyen.)

CHRYSOPHYLLUM CAINITO, Linn.

Star Apple.

Native of W. Indies and tropical America. A tree with leaves of a golden hue beneath; flowers purplish white; fruit size of an apple, green or purple. (*Sapotaceæ.*)

Fruit edible. Unripe—the juice with orange juice—astringent.

CINCHONA OFFICINALIS, Linn.

Cinchona, Crown Bark.

Native of the Andes of tropical S. America. A tree, often only a shrub, with opposite, simple leaves, rose-coloured, sweet-scented flowers, and small, winged seeds. (*Rubiaceæ.*)

Bark of root and stem contains four principal alkaloids,—quinine, quinidine, cinchonine, and cinchonidine. There are also acids and other substances.

"The Cinchona barks possess powerful antiperiodic, tonic, antiseptic, and astringent properties. The three first qualities are especially due to their alkaloids; and the latter to cincho-tannic acid and cinchona-red. The essential difference, therefore, between the action of the alkaloids and their salts, and of Cinchona barks, rests in the astringency of the latter. The alkaloids cannot therefore, in all cases, be substituted for the barks, for it is a well-ascertained fact, that there are cases which can be successfully treated by the preparation of the barks, when the salts of the alkaloids have failed to afford relief. This difference of effect is doubtless due in some instances to the astringent properties of the barks; but also in others, to causes not yet explained. Again, although the barks are applicable in all cases when the alkaloids have been found useful, they are apt when administered in large doses, to produce severe irritation of the stomach and bowels, and thus to cause nausea, or even vomiting and purging; they are also liable to cause constipation by their astringency. Hence, as antiperiodics, and in other cases, where large doses of the preparatiens of the barks are required to produce the desired effects, the salts of quinia, or of the other alkaloids, are, as a general rule, much better adapted for use. As tonics, the barks in small doses, are extensively employed in cases of debility, more especially if this be caused or accompanied by profuse discharges, as colliquative sweating or diarrhœa, or by other mucous or purulent discharges, as leucorrhœa, abscesses, &c. Garrod says that, 'as a tonic, in cases of great debility with weak heart, bark is more agreeoble and beneficial than quinine'; and also, that 'the astringent property of bark renders it better fitted for the treatment of relaxed conditions of the habit than quinine.' The preparations of the bark are also most useful after surgical operations when the strength has been greatly reduced; and in all cases of physical exhaustion, as in convalescence after either acute or lingering chronic diseases, unless there be much irritability of the stomach or bowels, or bowels, or inflammation, or febrile symptoms, when their use is contra-indicated." (Bentley & Trimen.)

Mr. J. J. Bowrey, Island Chemist, has made preparations from the bark grown in the island, both in the form of a dry powder, and a liquid. These preparations, while combining all the properties of the bark, obviate the necessity of the large doses.

Other species are *C. Calisaya*, Wedd., and its variety *Ledgeriana*, *C. succirubra*, Pav. (Red or Druggists' Bark), and a hybrid of *C. officinalis* and *C. succirubra*.

CINNAMODENDRON CORTICOSUM, Miers.

RED CANELLA, MOUNTAIN CINNAMON.

Native of Jamaica. A tree with lance-shaped leaves, and small scarlet flowers. Belongs to the same order as Canella. *(Canellaceæ.)*

BARK used formerly as a substitute for Winter's Bark, considered stimulant, tonic, aromatic, and antiscorbutic.

CINNAMOMUM ZEYLANICUM, Breyn.

CINNAMON.

Native of E. Indies, cultivated. A tree; leaves simple, opposite, with 3 to 5 strong nerves; flowers small. *(Laurineæ.)*

The plants are pruned to prevent trees being formed, four or five shoots are allowed. When the shoot turns brown in 18 months or 2 years, the bark is stripped off, and dried for the cinnamon of commerce.

LEAVES yield "clove oil."

BARK: Cinnamon is chiefly used as a spice.

Medicinally, it is aromatic, stimulant, carminative, and somewhat astringent.

Volatile oil possesses same properties without astringency Used also locally in paralysis of tongue, neuralgic headache, &c.; also in perfumery.

The bark yields this oil by distillation to extent of ½ and 1 per cent.

CINNAMON. See CINNAMOMUM ZEYLANICUM.

CINNAMON, Mountain. See CINNAMODENDRON CORTICOSUM.

CINNAMON, Wild. See CANELLA ALBA, & PIMENTA ACRIS.

CIRCASSIAN SEEDS. See ADENANTHERA PAVONINA.

CISSAMPELOS PAREIRA, Linn.

VELVET LEAF.

W. Indies, Central America, and extending through the tropics.

A woody climbing plant with rounded leaves, and minute, inconspicuous flowers. *(Menispermaceæ.)*

This plant was at one time supposed to be the source of Pareira Brava, and though it is not of equal value, it contains the same bitter principle.

ROOT "has the character of being diuretic and alterative. It is prescribed in dropsy, dysury, urinary calculus, jaundice, gout and cutaneous diseases. The infusion is recommended to be drunk freely during the irritable stage of gonorrhœa." (Macfadyen.)

LEAVES beat up into a pulp, applied to sores.

CISSUS SICYOIDES, Linn.

WEST INDIAN BRYONY.

Native of W. Indies and tropical America.

A shrubby climber, with tendrils; leaves simple; flowers small, greenish-yellow; berries black. *(Ampelideæ.)*

"It is used as an application to sores, and as a substitute for adhesive plaster.

LEAVES leave a pungent taste. When bruised in water, they make a lather like soap." (Macfadyen.)

CITHAREXYLUM SURRECTUM, Gr.

FIDDLE WOOD.

Native of Jamaica and Antigua. A tree; leaves simple; flowers white, fragrant; berry black. *(Verbenaceæ.)*

C. quadrangulare, Jacq. is very like this tree, and is also known as Fiddle Wood.

WOOD: "A most useful timber in building, close grained and very tough, used for mill rollers and frames, carriage wheels, &c." (Harrison.)

CITRON. See CITRUS MEDICA.

CITRUS AURANTIUM, Linn.

SWEET ORANGE.

Native of Northern India. A small tree; leaves with a winged stalk; flowers, white, fragrant; fruit with sweet juice. *(Rutaceæ.)*

LEAVES and young shoots yield by distillation a volatile oil, used in preparation of Eau de Cologne, and other perfumes. The oil from the Bitter Orange is of greater value.

FLOWERS, distilled with water, yield Orange Flower Water and Oil of Neroli. The "water" is a slight nervous stimulant and is often prepared by infusing the flowers. The water is chiefly used, however, as a flavouring agent

Oil of Neroli used in preparation of Eau de Cologne, perfumery, and liqueurs.

FRUIT, immature, as they drop from the tree, about the size of a cherry, yield the same oil. They are used to flavour Curaçoa, &c.

The ripe orange is highly valued as a fruit.

It is a refrigerant in fever and inflammatory complaints.

The PEEL has aromatic stimulant properties, and is slightly tonic, but the Bitter Orange is preferred.

An essential oil is premared from the peel, known as Essence de Portugal.

CITRUS AURANTIUM, var. BERGAMIA, W. & A.

WEST INDIA LIME.

Native of Northern India. A smaller tree, with smaller leaves, flowers and fruit, those in the Orange. *(Rutaceæ.)*

FRUIT yields a juice, which like that from the lemon is refrigerant and antiscorbutic, the latter property being due, according to Garrod, not to the citric acid, but to the potash salts contained in the juice. The juice is also given to counteract the effect of narcotic poisons.

Oil of Bergamot, used in perfumery, is prepared by placing the "full" fruit in a special machine, which ruptures the oil vessels in the rind.

The juice of the fruit is used with soap, &c., and the rind is preserved.

CITRUS AURANTIUM, var. BIGARADIA, Hook, fil.

BITTER ORANGE, SEVILLE ORANGE, BIGARADE ORANGE.

Native of Northern India. A small tree, the wing of the leaf-stalk broader than that in the Sweet Orange: fruit with bitter rind and pulp. (*Rutaceæ.*)

Orange Flower Water, Oil of Neroli, peel of fruit, oil from peel (called Essence de Bigarade), are all prepared from the Bitter Orange rather than from the Sweet Orange, *Citrus Aurantium*, which see.

Rind used for Marmalade.

CITRUS MEDICA, Linn.

CITRON.

Native of Northern India. A small tree; leaves with a short stalk and not winged; flowers pinkish, with purplish buds; fruit with transverse and longitudinal furrows, very thick, hard rind, pulp scanty with acid juice. (*Rutaceæ.*)

LEAVES and FLOWERS yield fragrant oils.

FRUIT—Rind yields an essential oil—Essence of Cedrat, used for perfumery.

Rind of fruit is candied, and used for dessert and in confectionery.

CITRUS MEDICA, var. LIMONUM, Hook. fil.

LEMON.

Native of Northern India. A small tree; leaves with a stalk which is not winged or only slightly so; flowers tinged with purplish-pink, with a fragrance distinct from that of Orange: fruit with smooth rind and acid pulp. (*Rutaceæ.*)

FRUIT—Outer rind is an aromatic stomachic.

Oil or essence of lemon (Essence de Citron) is obtained from the rind. It is a stimulant and carminative, and is used in perfumery and confectionery.

Juice used to prepare citric acid.

The juice is used in scurvy, and to counteract narcotic poisoning.

CLEMATIS DIOICA, Linn.

TRAVELLER'S JOY.

Native of W. Indies and tropical America. This native Clematis is a climber, with ternate leaves, greenish-white flowers, and the numerous seed-vessels terminating in a long feathery tail. (*Ranunculaceæ.*)

ROOT—"A decoction in sea-water, mixed with wine, is said to act as a powerful purge in hydropic cases." (Macfadyen.)

STEMS—Used as withes for tying.

LEAVES—Bruised into pulp, act as a rubefacient, and even vesicate.

LEAVES and FLOWERS—An infusion of these bruised used as lotion to remove spots from skin.

CLEOME PENTAPHYLLA, Linn.

BASTARD MUSTARD.

Found in all warm countries. An annual herb, with whitish flowers, and leaves with about 5 segments. *(Capparideæ.)*

"The juice, either plain, or mixed with sweet oil, is a certain remedy for the ear-ache. It ought to be warmed previous to being used. A preparation may be made by beating up the young branches of the plant, with sweet oil, in a mortar." (Macfadyen.)

CLETHRA TINIFOLIA, Sw.

SOAPWOOD.

Native of W. Indies, and tropical America. A tree; leaves simple; hairy on the under surface; flowers white. *(Ericaceæ.)*

"At the highest levels of the Blue Mountains. A small but useful timber." (Hooper.)

CLITORIA TERNATEA, Linn.

PEA FLOWER.

Tropics. A twining plant: leaves compound; flowers large, purple or white; pod flat. *(Leguminosæ.)*

ROOT—Powerful cathartic.

SEEDS, powdered more useful—purgative and aperient. Said to be a safe medicine, especially for children.

CLOVE. See CARYOPHYLLUS AROMATICUS.

CLOVE, WILD. See PIMENTA ACRIS.

CLUB MOSS. See LYCOPODIUM CLAVATUM.

COCA. See ERYTHROXYLON COCA.

COCCOLOBA UVIFERA, Jacq.

SEASIDE GRAPE.

Native of W. Indies and tropical America.

A tree, with roundish cordate leaves; flowers without petals, and hanging bunches of dark-blue berries. *(Polygonaceæ.)*

WOOD: A crooked tree, wood hard and takes a fine polish, used for fancy work. (Harrison.)

"On the coast line inside Mangrove In Jamaica this species remains small, whereas elsewhere it grows into a large tree, notably in Honduras. The wood is hard and takes a polish." (Hooper.)

FRUIT: So very astringent that caution has to be exercised in its use.

COCCULUS INDICUS. See ANAMIRTA COCCULUS.

COCCUS WOOD. See BRYA EBENUS.

COCO. See COLOCASIA ANTIQUORUM.

COCOA. See THEOBROMA CACAO.

COCOA-PLUM. See CHRYSOBALANUS ICACO.

COCO-NUT PALM. See COCOS NUCIFERA.

COCHINEAL CACTUS. See OPUNTIA COCCINELLIFERA.

COCOS NUCIFERA, Linn.

COCO-NUT PALM.

Tropics. A palm with pinnate leaves. (*Palmæ.*)

"Toddy" is obtained from the flower spathe just before it opens by slicing off the top, and collecting the sap in a vessel. It has a pleasant, sweetish taste, and in large doses is aperient. Fermented it is intoxicating. It can also be boiled down into a course sugar called "jaggery," which is refined, or fermented and distilled into spirits.

The young Coco-nut contains a sweet refreshing water and jelly.

The nut is generally harvested before it is perfectly mature. If the outer skin dries on the tree the fibre of the husk becomes coarse and dark in colour; if too young, it is weak.

Coco-nut milk is made from gratings of the kernel. The shell is carved and used for many purposes.

The dried kernel is known as "kopra," and is used for the preparation of oil by expression or boiling. The solid fat is employed in making candles, and the oil for cooking, for lamps, as a substitute for cod-liver oil, &c. The cake which is left, or "poonac," is a good food for cattle and is also used as a manure.

The husk of the fruit yields Coir-Fibre. "Coir is remarkable for its durability, and is used for the manufacture of various textile fabrics, brushes, cordage for the rigging of ships, nets, matting, stuffing of cushions, pads and mattresses, scrubbing brushes, fishing net, &c. The tender leaves are used for plutting mats, boxes, and other fancy articles. The mature leaves are plaited into matting, and also used as materials for fences, sails, buckets, books, fans, torches, and fuel The ash yields an abundance of potash. The midribs of the leaflets are made use of as brooms, brushes, and skewers. The stalk of the spadix itself is in every day use as a chunam brush to whitewash houses with. The reticulated web of the base of the leaf forms a coarse kind of cloth. The cottony hairs are used as a styptic. The soft parts within the stem of the Coco-nut are cut out and pounded in a mortar; the resulting pulp is washed in water, and the farina is collected, and used as a substitute for sago.

Aged and unfruitful trees are cut down, and the wood is turned to a variety of useful purposes; is is hard, handsome, and durable, known under the name of Porcupine Wood: it is used for veneering. The hard stem is converted into drums, gutters, water-pipes, small boats, frames, furniture, rafters for houses, spear-shafts, shingles, walking-sticks, ladies' work boxes, &c. The root stem takes a high polish so as to resemble agate. A cubic foot weights 70 pounds and the wood is supposed to last 50 years." (Dr. John Shortt.)

A dye can be extracted from every part of the plant, producing a dirty-brown colour. Gum is said to be yielded in Tahiti.

COFFEA ARABICA, Linn.

COFFEE.

Native of tropical Africa. A large shrub, with opposite leaves, white flowers, and scarlet berries. (*Rubiaceæ.*)

BERRIES contain caffein; roasted, they develop also a brown bitter principle and a volatile oil. Coffee is an agreeable, stimulating, soothing, and refreshing beverage. See Bulletins, Nos. 4, 5, 6, 8, 12, 14.

Liberian Coffee (*C. liberica, Hiern*) has a larger berry with a hard, fibrous pulp. It does not appear to be so liable to disease as common coffee, and endures tropical heat better.

COGWOOD. See ZIZYPHUS CHLOROXYLON.

COIX LACHRYMA, Linn.

JOB'S TEARS.

Native of India. Cultivated in gardens in Jamaica. A tall grass. (*Gramineæ.*)

GRAIN used by some hill tribes in India as food, but chiefly used for ornamental purposes.

COLA ACUMINATA, R. Br.

BISSY, KOLA, COLA.

Native of tropical W. Africa. A tree of moderate height, leaves; simple, 6 or 8 inches long; flowers with a pale-yellow calyx, but without petals; seeds large.

SEEDS used like chocolate. They are said to be of great dietetic value, and also to be useful in dyspepsia and nervous diseases. "It has been said that the beverage made with Cola paste is ten times more nutritious than chocolate made with cocoa. The reputation of this substance in sustaining the system against fatigue is such that it is meeting with consideration from the military authorities of the world as an article to be given to soldiers during active service." (Watt.) Seeds have been sold lately in London at 2s. and even 3s. per lb.

COLOCASIA ANTIQUORUM, Schott.

COCO, EDDOES.

Native of East Indies. An arum-like plant, with large heart-shaped leaves, and tuberous root-stocks. (*Aroideæ.*)

LEAVES—"Juice expressed from the leaf stalks of the black species is used with salt as an absorbant in cases of inflamed glands and buboes. The juice of the tuber is used in cases of alopecia. Internally, it acts as a laxative, and is used in cases of piles and congestion of the portal system, also as an antidote to the stings of wasps and other insects . . . Have seen remarkable instances of its styptic properties; if applied to fresh and clear wounds, it enables the tissues to unite by first intuition within a few hours." (Watt.) Young leaves may be eaten like spinach.

ROOTS.—Tubers often weigh several pounds, and form a nutritious food when well cooked.

COMOCLADIA INTEGRIFOLIA. Jacq.

Maiden Plum.

Native of Jamaica. A small tree, 10 to 30 feet high; leaves pinnate, at the top of the stem or branchlet; flowers very small, purple. *(Anacardiaceæ.)*

Wood.—A very hard wood, does not grow large enough for sawing, but makes an almost everlasting post for which purpose it is much used. (Harrison.)

"Found in most places, except at high elevations. A small tree, in hedges used for posts, which grow when planted. The timber is said to be very hard." (Hooper.)

Wood is also said to afford a red dye. The juice of the whole plant is an indelible marking ink.

CONGO PEA. See Cajanus indicus.

CONOCARPUS ERECTA, Linn.

Button Wood.

Native of W. Indies, and tropical America; also found in west tropical Africa. A small tree, with alternate, simple leaves; flowers without petals, clustered in heads. *(Combretaceæ.)*

Wood—Not large, used for piles, and stands well in salt water. (Harrison.)

"Along the coast, notably St. Elizabeth, Westmoreland. Growing to no great dimensions, but very useful for posts, being lasting both in and out of the ground." (Hooper.)

CORATOE. See Agave Morrisii.

CORDIA GERASCANTHOIDES, H. B. K.

Spanish Elm.

Native of Jamaica, Cuba and Mexico.

A tree, 20-30 feet high, a foot in diameter at the base; with single leaves, and white flowers covering the tree. *(Boragineæ.)*

"This tree makes a good post to go in the ground, and is much used by the peasantry in the erection of their houses. Hoops and staves are also made from it." (Harrison.)

"At the base of the hills in Clarendon, St. Thomas-in-the East, Portland and elsewhere. No specimens of any size to be found now. It is useful in underground work and coopering." (Hooper.)

CORN. See Zea Mays.

CORK WOOD. See Anona palustris and Ochroma Lagopus.

COTTON. (See Gossypium Barbadense.)

COTTON, FRENCH. See Calotropis procera.

COWHAGE See MUCUNA PRURIENS.

COWITCH. See MUCUNA PRURIENS.

CRABS EYES. See ABRUS PRECATORIUS.

CRESCENTIA CUJETE, Linn.

CALABASH.

Native of W. Indies and tropical America. A tree; leaves narrow, 4-8 inches long; flowers 2 to 3 inches long, variegated in colour; fruit pulpy with a hard shell.

WOOD is tough, light and elastic, but as it is of a crooked growth, it is only used for small work, such as the felloes of wheels and in carriage building generally, cattle yokes and a variety of small articles.

The shell of the fruit makes excellent drinking cups, vessels for carrying water, dish covers, &c.

FRUIT.—Expressed juice of the pulp—purgative; also a demulcent and pectoral.

CROTALARIA JUNCEA, Linn.

SUNN HEMP, BOMBAY HEMP.

Native of India, Malay Islands, and Australia. An annual shrubby plant, belonging to the Pea Family, (*Leguminosæ,*) of erect habit, growing sometimes as high as 10 feet, with bright yellow flowers. It is naturalised in Jamaica, but is not anything like as common as *Crotalaria retusa*, which is cultivated for fibre in Madras.

The soil must be rich and friable. To obtain stems without branches for fibre, the seed is sown close. The plant is sometimes grown for fodder, especially for milch cows, and then seed is sown at greater intervals. The plants are ready for harvesting in 4 or 5 months. If a soft fibre is wanted, the plants are pulled in flower; if a strong fibre is desired the plants are left until the seeds are almost ripe. Retting is necessary and takes 3 days. The stems are then bent so as to break the wood, and they are beaten on the surface of the water, until the fibre comes away. It is hung up to dry, and finally combed out. The fibre is used for cordage, coarse cloth, and the waste fibre for paper.

CUCUMBER, WILD. See CUCUMIS ANGURIA.

CUCUMIS ANGURIA, Linn.

WILD CUCUMBER.

Native of West Indies and tropical America.

A climbing plant, with tendrils; leaves five-lobed; flowers yellow, small; fruit spiny. (*Cucurbitaceæ.*)

FRUIT.—A wholesome vegetable, also used in pickles.

CUPANIA EDULIS, Camb.

AKEE.

Native of tropical West Africa. Naturalized in Jamaica. A tree;

leaves pinnate; flowers white; fruit red, splitting on the tree, displaying the black seeds half enclosed with a white covering. (*Sapindaceæ.*)

Seeds.—The white covering is a wholesome vegetable.

CUSTARD APPLE. See Anona reticulata.

CUTCH. See Acacia Catechu.

CYPERUS ARTICULATUS, Linn.

Adrue.

Tropics. A sedge, with a small leaf; stem cylindrical, 2 to 4 feet high; root-stock knobbed. (*Cyperaceæ.*)

Root-Stock. "Aromatic and stimulant, may be used in the place of Virginia Snake-root,—infusion good in vomitings, fluxes, &c."

CYPERUS ROTUNDUS, Linn.

Nut Grass.

Tropics, and warmer temperate countries. A sedge, with flat leaves; stem 3-cornered, about 1 foot high; rootstock bearing tubers. A troublesome weed in gardens.

Tubers yield an essential oil, used in perfumery. The natives of India use the dried and pounded tubers as a perfume.

Used medicinally as a diaphoretic and astringent. Stimulant and diuretic properties are also attributed to them. They are further described as vermifuge. In native practice in India, they are held in great esteem as a cure for disorders of the stomach and irritation of the bowels. The bulbous roots are scraped and pounded with green ginger and in this form mixed with honey they are given in cases of dysentery in doses of about a scruple. It was well known to the Greeks and Romans, and is mentioned by Homer. (Watt.)

DATE PLUM. See Diospyros tetrasperma.

DATURA STRAMONIUM, Linn.

Thorn Apple. Stramonium.

Found in temperate and warm climates.

A weedy annual, 3 feet high, with large leaves, indented and irregular; flowers white, funnel-shaped; fruit leathery, set with spines. (*Solanaceæ.*)

Datura Tatula, Linn., with purple stems and flowers, is only considered a variety by Hooker.

Leaves and Seeds have the same properties, though the seeds are more active. Action seems to be the same as belladonna.

Properties, anodyne and antispasmodic. Useful in neuralgic and rheumatic affections, in gastrodynia, spasmodic asthma. In overdoses it is a powerful poison. Used as ointment, fomentation, &c., to allay pain in tumours, rheumatism, &c.

DIOSPYROS TETRASPERMA, Sw.

PIGEON WOOD. DATE PLUM.

Native of Jamaica, Cuba and Panama. A shrub, leaves simple; flowers yellowish; berry ½ inch diam. (*Ebenaceæ.*)

WOOD.—"This is a very good timber wood, lasts in the ground well, it is used for posts, scantling, &c. Wild pigeons feed on the berry, hence the name." (Harrison.)

"Found on the southern coast ranges and in the valleys. Of no great size. Gives a good timber." (Hooper.)

DIPHOLIS MONTANA. Gr.

BULLET OR BULLY TREE.

Native of Jamaica and Cuba. A tree; leaves simple, arch-veined; flowers small; berry about ½ inch. (*Sapotaceæ.*)

WOOD—"This is an excellent building timber and is much used sawn into boards, planks and scantlings. It is very durable and very lasting; shingles are made from it. It is found all over the Island, and grows 3 to 4 feet in diameter." (Harrison)

D. montana Gr., is the Mountain Bully Tree; *D. nigra*, Gr., is the Black or Red or Bastard Bully Tree; *D. salicifolia*, A. DC., is the White Bully Tree, also called Galimeta Wood, or Pigeon Wood.

"General over the Island below 3,500 feet. These together form an important class of timbers, and the trees, though vulgarly classed as bullet or bully trees, are not all belonging to one species. They give hard, heavy and close-grained timbers, used largely in general construction, some being valuable from being impervious to rot, either under water or above ground; others, notably the white bullet, are split into shingles. These bullet trees demand further study." (Hooper.)

DIVI-DIVI. See CÆSALPINIA CORIARIA.

DOG WOOD. See PISCIDIA ERYTHRINA.

DOLICHOS TUBEROSUS, Lam.

YAM BEAN.

Native of tropical Asia. A twining plant: leaves compound; flowers white; pod straight, compressed, with reddish hairs; seeds red; root formed of a number of fibres, several feet long, bearing tubers.

ROOTS "afford a plentiful supply of wholesome food. The produce of 3 plants is usually a bushel. The tubers may either be boiled plain, in which state they are a very good substitute for yams or other roots in common use; or they may be submitted to a process similar to arrow-root, and a starch obtained. The starch is of a pure white, and is equal in every respect to arrow-root. To the taste it is very palatable, is easily digested, and is employed for custards and puddings. Even the trash, left after obtaining the starch, and which in the preparation of arrow-root is lost, may, when thoroughly dried, be formed into a palatable and wholesome flour. A very excellent flour may also be obtained by slicing the tubers, drying them in the sun, and then reducing to a powder. This plant is deserving of being more generally cultivated than it has been. It can be planted at any season of the year, and the roots are fit for digging in the course of 4 or 5 months; the return is infinitely greater than that from arrow-root, and the proportion of starch also is more abundant." (*Macfadyen.*)

The young pods may be used like French Beans, but the ripe beans are poisonous.

DOWN TREE. See OCHROMA LAGOPUS.

DUTCHMAN'S LAUDANUM. See PASSIFLORA MURUCUJA.

EARTH NUT See ARACHIS HYPOGÆA.

EBONY, WEST INDIAN. See BRYA EBENUS.

EDDOES. See COLOCASIA ANTIQUORUM.

ELM, SPANISH. See CORDIA GERASCANTHOIDES.

ERIGERON CANADENSE, Linn.

HORSE WEED, BUTTER WEED.

Native of America and W. Indies. An annual herb, 2-4 feet high, of agreeable, aromatic odour, with minute flowers of very pale violet-white colour, in numerous small heads. (*Compositæ.*)

The plant should be collected, for medicinal use, while in flower. The leaves and flowers are most active. Its properties are imparted both to alcohol and water, but they are injured by boiling in consequence of the loss of volatile oil.

A tonic, astringent, and diuretic. Useful in diarrhœa, dysentery, and dropsical complaints.

"The oil was first introduced into notice by the Eclectic Physicians, who had found it useful in diarrhœa and various hæmorrhages. More recent trials seem to indicate that it is a remedy of more especial value in uterine hæmorrhage. The dose is from 5 to 10 drops every two hours. It has also been recommended in doses of ten drops in gonorrhœa. It is also employed by the eclectic practitioners when dissolved in alcohol, in the proportion of one drachm of the oil to from one to two ounces of alcohol, as an application to inflamed and enlarged tonsils, in inflammation and ulceration of the throat generally, and also in other local inflammations. It is said to be a most valuable remedy in such cases." (Bentley and Trimen.)

ERIODENDRON ANFRACTUOSUM, DC.

SILK-COTTON TREE, CEIBA.

Native of W. Indies and tropical S. America. A very high tree; leaves compound, palmate; flowers rose-coloured 1½ inch long; seeds many, enveloped in wool. (*Malvaceæ.*)

"Dotted here and there over the island up to 3,000 feet. A very large spreading tree, growing in the open, with peculiar twisted far-extending buttresses. The girth of the cotton tree is up to 30 feet, the timber soft and useless, except for cheap canoes, which last one season." (Hooper.)

"The large stems of this tree are hollowed out to form canoes. The wood is soft and subject to the attack of insects; but if steeped in strong lime water, it will last for several years, even when made into boards or shingles, and in situations exposed to the influence of the weather.

The wool has been employed in stuffing mattresses; and is said to answer the purpose equally well as feathers." (Macfadyen.)

ERYNGIUM FÆTIDUM, Linn.

Fit-weed.

Native of W. Indies and tropical America. A fetid herb, about a foot high; flowers small, white, collected in spiny heads; leaves spiny.

This plant has the character of being aphrodisiac, alexipharmic, and emmenagogue, and of being serviceable in colic, hysteria, and spasmodic diseases in general. It may be used in the form of infusion or decoction, or the root may be given in powder. It has received its common name from its efficacy in nervous diseases. (Macfadyen).

ERYTHROXYLUM AREOLATUM, Linn.

Redwood.

Native of Jamaica, Venezuela and New Granada. A shrubby tree, 10-16 feet high; leaves marked as in E. Coca with lines parallel with the midrab, 1½-2 inches long; flowers appearing before the leaves, white and fragrant; berries numerous, bright scarlet. (*Linaceæ.*)

Leaves contain only a small amount of Cocaine.

Wood.—"This timber grows to medium size, saws readily, not too hard for general purposes, is used for furniture and flooring." (Harrison.)

ERYTHROXYLON COCA, Lam.

Coca Shrub.

Native of the Andes. Cultivated in Jamaica. A shrub with pale yellow flowers, and red berries. The leaves are unmistakable from the two curved lines on the under surface, one on either side of the midrib. (*Linaceæ.*)

Leaves dried, form the coca of commerce. The chief constituents of coca are a crystalline alkaloid, called *cocaine*, and a volatile odoriferous alkaloid, *hygrine*. See Bulletins, 15, 16.

ESPARTO, See Stipa tenacissima.

EUCALYPTUS GLOBULUS, Labill.

Blue Gum

Native of Australia and Tasmania. Established at Cinchona, Jamaica. A tree of very rapid growth, and in Australia, attaining sometimes the extraordinary height of over 300 feet. Leaves on young shoots are opposite, ovate, of a very pale greenish-blue colour; on old branches the leaves are not opposite, they are sabre-shaped, with the stalk twisted so that they hang vertically. (*Myrtaceae.*)

Leaves contain numerous oil-glands, from which a volatile oil is obtained by distillation.

It is to this oil that the fibrifugal properties of the plant are due. The alcholic tincture is the best form in which to administer. It has been successfully used in ague, periodic fever, palustral cachexia, ailments of an atonic or anæmic character, as a stimulant and antispos-modic, and in bronchitis. The leaves are used to dress wounds.

"It can scarcely be doubted that this tree does produces a most beneficial effect by destroying the fever-producing miasm of marshy districts." (Bentley and Trimen.)

This species is not suited to low elevations in Jamaica, but experiments are being made to find species that will stand the climate, and at the same time act beneficially in malarious districts.

EUPATORIUM NERVOSUM, Sw.

Bitter Bush.

Native of Jamaica and Haiti. A perennial herb, 4 or 5 feet high; leaves opposite, ovate, 1 to 3 inches long, dotted beneath with minute glands; flowers whitish. (*Compositae.*)

An infusion of the leaves and tops, gathered after flowering has commenced, is "regarded as efficacious in cholera, and also in typhus and typhoid fevers, and in small-pox: it is also reputed to be a good cholagogue." (Bentley & Trimen.)

EUPHORBIA PILULIFERA, Linn.

Tropics. An annual weed: leaves simple, pointed, ½ to 1 inch long; flowers minute; seeds 4-cornered. (*Euphorbiaceæ.*)

The whole plant is useful "in cases of asthma and bronchitis, to relieve spasm, and promote free expectoration." (Dr. Henderson.)

FEVER GRASS. See Andropogon citratus.

FIDDLE WOOD. Vitex umbrosa.

FORSTERONIA FLORIBUNDA, G. Don.

Milk Withe.

Native of Jamaica. A climbing shrub, leaves simple, 2 to 3 inches long; flowers small, whitish; seeds nearly ½ inch long with brownish hairs. (*Apocyneæ.*)

Stem yields caoutchouc, valued at 3s. 2d. per lb. See Bulletins x, and xxi.

FEVILLEA CORDIFOLIA, Sw.

ANTIDOTE CACOON.

A climbing plant, with tendrils; leaves roundish, 3-4 inches; flowers small, orange colour; fruit, size of an apple of a russet colour, hard, full of large, flat, round seeds. *(Cucurbitaceæ.)*

SEEDS abound in oil; a good torch can be made by stringing them on a thin stick. Oil "has been manufactured into candles."

"The Spanish physicians, we are told, employ the seeds with success in the form of an emulsion, for intermittent fever, and as a counter-poison. The Buccaneers esteemed it so highly, that they never ventured on an expedition without taking with them a supply of this fruit." *(Macfadyen.)*

FIDDLE WOOD. See CITHAREXYLUM.

FIT WEED. See ERYNGIUM FŒTIDUM.

FLAX, NEW ZEALAND. See PHORMIUM TENAX.

FRENCH COTTON. See CALOTROPIS PROCERA.

FRENCH OAK. See CATALPA LONGISSIMA.

FURCRŒA CUBENSIS, Haw.

SILK GRASS.

Native of West Indies and tropical America.

A plant like an Agave, with spiny leaves, white flowers and very short stem. *(Amaryllideæ.)*

Leaves yield a fibre, which may supply a small part of the Sisal Hemp of Yucatan.

FURCRŒA GIGANTEA, Vent.

MAURITIUS HEMP.

Native of Central America. This plant is very much like the one known as Silk Grass, but it is larger and has a distinct stem.

It probably yields some of the fibre exported from Yucatan as Sisal Hemp, but it is not the true plant, and the price of the fibre is not as high. It was introduced many years ago into Mauritius, where it rapidly spread. When a demand arose for fibre there was an immense quantity in Mauritius ready at hand, and there was no expense incurred in planting.

FUSTIC. See MACLURA TINCTORIA.

GENIP. See MELICOCCA BIJUGA.

GINGER. See ZINGIBER OFFICINALE.

GOSSYPIUM BARBADENSE, Linn.

COTTON.

Native probably of W. Indies and tropical America. The mummy cloths of Peru are cotton, of Egypt linen.

A small shrub, with lobed leaves, yellow flowers, and seeds covered with long, white hairs. (*Malvaceæ.*)

BARK of root used as an emmenagogue, also in dysmenorrhœa.

SEEDS yield oil, which is used for various purposes, and sometimes as a substitute for olive oil.

The cake, left after the expression, is given to cattle.

A decoction of the seeds is a remedy in intermittents.

HAIRS from the seeds constitute commercial cotton, one of the most important fibrous materials.

Cotton consists of nearly pure cellulose, which has the same chemical formula as starch. By the action of nitric and sulphuric acids, it is converted into the explosive substance *Gun Cotton*, known in the Pharmacopœia as *Pyroxylin*. This dissolved in a mixture of ether and rectified spirit, gives *Collodion*. *Flexible Collodion* is made by adding a small quantity of Canada Balsam and Castor oil to collodion. Collodion is used to apply to wounds, skin diseases, &c. : the ether evaporates, leaving a thin film.

GROUND NUT. See ARACHIS HYPOGÆA.

GOUANIA DOMINGENSIS, Linn.

CHEW STICK.

Native of West Indies and tropical South America.

A shrubby climber, with tendrils; leaves simple: flowers small, yellowish, in clusters; seed-vessel 3-winged. (*Rhamneæ.*)

This is a very agreeable bitter. It is used as a substitute for hops in ginger beer, and cool drinks.

The infusion has been employed in gonorrhœa and dropsy, and as a light grateful bitter, in cases of debility, to restore the tone of the stomach.

In powder, it forms an excellent dentifrice; its aromatic bitter producing a healthy state of the gums, and the mucilage it contains working up by the brush into a kind of soap-like froth. A tincture also is prepared from it, diluted with water, as a wash or gargle, in cases of salivation or disease of the gums. Chew-stick is also a substitute for the tooth-brush itself. A piece of a branch, about as thick as the little finger, is softened by chewing, and then rubbed against the teeth. In this manner a tooth-brush, and, with it, a powder are obtained, equal, if not superior, to any in use in Europe. (Macfadyen.)

GOURD, BOTTLE. See LAGENARIA VULGARIS.

GRANADILLA. See PASSIFLORA QUADRANGULARIS.

GRAPE, SEASIDE. See COCCOLOBA UVIFERA.

GRAPE VINE. See VITIS VINIFERA.

GRAPE, WILD. See *Vitis caribæa.*

GREENHEART. See SLOANEA JAMAICENSIS.

GUAIACUM OFFICINALE, Linn.

LIGNUM VITÆ.

Native of West Indies, Venezuela, Colombia. A small tree; leaves compound, leaflets opposite in 2 or 3 pairs; flowers of a bright pale blue, covering the tree: berries brownish-yellow. (*Zygophyllaceæ.*)

WOOD:—hard, tough, dense, and durable. Used for pulleys, blocks, pestles, rulers, skittle balls, &c.

"On the alluvial deposits near the coast. A tree up to two feet six inches in diameter (Mitchelltown, Vere), though rarely exceeding one foot and a half, with a height of 15 feet. A hard, close-grained wood, exported to Europe for the manufacture of small articles. Since 1865 5,380 tons have been exported, valued at £12,829." (Hooper.)

The heart wood is of a dark greenish-brown colour, owing to the deposition of guiacum resin; the sap wood is pale-yellow.

The wood is official in the Pharmacopœias, and owes its medicinal properties to the presence of the resin, which is also official. "Guiacum resin possesses stimulant, diaphoretic, and alterative properties like the wood; but its action is much stronger. By some practitioners it is also regarded as an emmenagogue. It is a useful remedy in chronic forms of rheumatism; also in syphilitic and gouty affections, scrofula, skin diseases; and in dysmenorrhœa, and other uterine affections, &c." (Bentley & Trimen.)

GUANGO. See PITHECOLOBIUM SAMAN.

GUAREA SWARTZII, D. C.

ALLIGATOR WOOD, MUSK WOOD.

Native of Jamaica and Guadeloupe.

A low tree; leaves pinnate, flowers small, white. (*Meliaceæ.*)

"Common on Blue Mountains. Used for shingles." (Hooper.)

"All parts of the tree especially the bark, have a strong smell of musk, resembling that of the Alligator. There cannot be a doubt, that many parts of this tree are possessed of medical properties. The powdered bark is a good emetic. The seeds are bitter, and have a warm musky taste." (Macfadyen.)

GUAVA. SEE PSIDIUM GUAVA.

GUAVA, MOUNTAIN. SEE PSIDIUM MONTANUM.

GUAZUMA TOMENTOSA, H. B. K.

BASTARD CEDAR.

Native of West Indies and tropical America.

A tree 15 to 50 feet high; leaves, simple, in two rows on the twigs; flowers small, yellow; nut purplish-black.

WOOD; light and splits readily.

"Found in the open forest in the vicinity of high roads and habitations. This is a valuable tree, mainly from its yielding foliage and fruit which are readily eaten by stock of all kinds. It is of a spreading habit, rarely taller that 20 feet, with a diameter at the base of three feet; it yields a good timber, but it is rarely cut until it ceases to bear fruit." (Hooper.)

BARK: "an infusion has been employed medicinally, and given internally as a remedy for coco-bay, elephantiasis, and other obstinate cutaneous diseases." (Macfadyen.)

HAEMATOXYLON CAMPECHIANUM, Linn.

LOGWOOD.

Native of tropical America. A tree, with pinnate leaves, small yellow flowers, and small pointed pods. *(Leguminosae.)*

WOOD.—The heart-wood is red-coloured, and this alone is exported, the whitish sapwood being chipped off. The root is also exported. The wood is employed to produce violet and blue colours, shades of grey, and more especially blacks, giving to the latter a velvety lustre. "On the coast, but especially on the east, south-east and south. In Clarendon and St. Elizabeth it has spread inland up to 2,000 feet altitude. Introduced from British Honduras in 1715, though it is believed an entrepot trade in it had been carried on previously, shipments from Honduras being landed and re-shipped from Jamaica. It generally acquires in from 10 to 20 years a diameter of 9 inches to a foot, with a stem branching upwards at a height of six feet from the ground. There are large trees in the Clarendon Hills. . . .

A clump of large trees exists on Goshen Common measuring up to 24 feet in girth. The yearly export varies from 22,000 to 114,900 tons. As a dye wood it is inferior to both Honduras and Campeachy wood, judging from the market prices, which are in the proportion of five, seven, and nine, the last being given for Campeachy wood. Of recent exports a large proportion is comprised of the roots of previous cuttings, removed because the logwood does not coppice." *(Hooper.)*

"Logwood is a mild astringent. It has been found useful in chronic diarrhœa and dysentery, in some forms of atonic dyspepsia, and especially in the diarrhœa of infants. As an injection the decoction of logwood has been found of service in leucorrhœa; and in the form of an ointment prepared from the extract of logwood, it is said to be useful in cancer and hospital gangrene." (Bentley & Trimen.)

Logwood makes a strong and durable fence, but must be kept well pruned.

HEART PEA. See CARDIOSPERMUM HALICACABUM.

HELICTERES JAMAICENSIS, Jacq.

JAMAICA SCREW TREE.

Native of West Indies and Central America.

A shrub or low tree, 4 to 15 feet high, with simple leaves, white flowers, and fruit shaped like a twisted cone. (*Malvaceæ.*)

" A decoction or infusion of the leaves and fruit may be used as a substitute for a similar preparation of marsh mallow, and given as a drink in fevers, consumption, cough, &c." (Macfadyen.)

HEMP. See CANNABIS SATIVA.

HEMP, BOMBAY. See CROTALARIA JUNCEA.

HEMP, MANILA. See MUSA TEXTILIS.

HEMP, SUNN. See CROTALARIA JUNCEA.

HEMP, MAURITIUS. See FURCRŒA GIGANTEA.

HIBISCUS ABELMOSCHUS, Linn.

MUSK OCHRA.

Found in all tropical countries. A somewhat shrubby plant, flowers sulphur-yellow with crimson base, seeds with a musky smell. (*Malvaceæ.*)

SEEDS sometimes used as a substitute for musk in perfumery; and in nervous and spasmodic diseases.

HIBISCUS CLYPEATUS, Linn.

CONGO MAHOE.

Native of Jamaica. A shrub, 6 to 12 feet high, with reddish-yellow flowers. (*Malvaceæ.*)

BARK—fibre used for cord and whip-lashes.

HIBISCUS ELATUS, Sw.

BLUE OR MOUNTAIN MAHOE, CUBA BARK.

Native of West Indies. A tree, 50 or 60 feet, with roundish leaves, large flowers of a purplish-saffron colour. (*Malvaceæ.*)

BARK—fibres make good ropes. The lace-like inner bark was at one time known as Cuba bark from its being used as the material for tying round bundles of Havanna cigars.

WOOD, valuable to cabinet-makers; best variety has the appearance of dark-green variegated marble.

LEAVES and young shoots, mucilaginous; infusion used in dysentery.

"This wood is much used in building, it makes a very pretty flooring and pretty furniture, picture frames, &c. When fully ripe, it is of a dark blackish-green colour, with darker or lighter bands, and makes a pretty contrast with lighter woods. In some localities boards 2 to 3 feet may be got. The bark yields an excellent fibre much used for cordage." (Harrison.)

"On the limestone at altitude of 2,000 feet and upwards a tall, stout tree, diameter up to three feet, height 80 feet. Much sought after for furniture and interior woodwork on account of its curious striped grain; also gives good shingles. Bark yields a fibre." (Hooper.)

HIBISCUS ESCULENTUS, Linn.

OCHRA.

Found in all tropical countries.

A large annual herb, flowers pale-yellow with a red base, fruit cylindrical, 3-6 inches long. (*Malvaceæ.*)

The WHOLE PLANT abounds in a viscid mucilage, and the unripe FRUIT is official in the Indian Pharmacopœia. In the form of a decoction, the fruit may be used in catarrhal affection, gonorrhœa, dysuria, &c. "The inhalation of the vapour of the hot decoction has been found very serviceable in allaying cough, hoarseness, irritation of the glottis, and other affections of the throat and fauces." (Waring.)

The principle use of ochra fruit is as a vegetable and to thicken soups, &c.

SEEDS yield an oil similar to olive oil.

STEM—a fibre of excellent quality is obtained, and a patent has taken out in France for making paper from it, called *banda* paper.

HIBISCUS SABDARIFFA, Linn.

RED SORREL.

Cultivated in tropics. An annual herb, with yellow corolla, and red stem, branches, &c. (*Malvaceæ.*)

ROOT—Gentle, laxative.

STEM yields fibre, which is fine and silky.

FLOWER—Outer envelopes made into preserves, tarts, and an infusion is a refreshing beverage, called Sorrel-drink. ("Rozelle" of India.)

HIBISCUS TILIACEUS, Linn.

SEA-SIDE MAHOE.

Native of tropics. A tree, 10 to 20 feet high, with roundish leaves and large yellow flowers. (*Malvaceæ.*)

BARK affords a strong fibre. Dampier often refitted the rigging of his ships with rope made from this bark.

HOG GUM. See SYMPHONIA GLOBULIFERA.

HOOP TREE. See MELIA SEMPERVIRENS.

HORSE EYE BEAN. See MUCUNA URENS.

HORSE RADISH TREE. See MORINGA PTERYGOSPERMA.

HORSE WEED. See ERIGERON CANADENSE.

HYMENÆA COURBARIL, Linn.

Locust Tree.

Native of West Indies and tropical America. A lofty spreading tree leaves compound with two leaflets; flowers white; pod with about three seeds enclosed in a mealy substance. (*Leguminosæ*.)

Roots—"A fine transparent resin of a yellowish or red colour exudes between the principal roots. It is the *Gum Animi* of the shops. It requires highly rectified spirits of wine to dissolve it, and makes the finest varnish that is known, superior even to the Chinese *lacca*. It burns readily, emitting a grateful and fragrant smell, and has been employed by way of fumigation in attacks of spasmodic asthma, and other embarrassments of respiration. In solution, it is given internally in doses of a teaspoonful, as a substitute for Gum Guiacum, for rheumatic and pseudo-syphilitic complaints, and employed externally as an embrocation. From this resin an oil may be distilled." (Macfadyen.)

Wood, takes a fine polish, adopted for making cogs of wheels in machinery. (Macfadyen.)

"Hard and heavy, very durable, saws easily, used in general building." (Harrison.)

"Near the coast, Liguanea Plain, up the valley of the Black River, and elsewhere. A large tree, with thick fleshy pod which smells offensively when opened, attains in the open a diameter of four feet and height of 25 feet (Mona carriage drive), elsewhere much taller. Wood described as hard and heavy in house building. In other West Indian islands much prized for furniture and cabinet work" (Hooper.)

Bark—Decoction of inner bark, vermifuge.

Pod—Mealy substance sweet and pleasant, eaten by Indians.

INDIGOFERA TINCTORIA, Linn.

Indigo.

Native of Asia and Africa. A small shrub, belonging to the Pea Family (*Leguminosæ*), with pinnate leaves, small pinkish flowers, and pod 1 to 1½ inch long. This plant is the one so largely cultivated in the East Indies, but another species, *Indigofera Anil*, Linn. (Wild Indigo) also yields the dye; it is a native of the W. Indies and tropical America, and it is readily distinguished from the other species by the pod which is only ½ inch long and much curved.

"For preparing indigo, the plants are cut down, just before flowering, placed in troughs, and after being pressed closely together they are covered with water. Fermentation takes place, and is allowed to continue from 12 to 15 hours, when the body of the liquid acquires a sherry colour, and the surface becomes covered with a blue film. It is then decanted, and the colouring principle dissolved by the water, absorbing oxygen from the air, becomes insoluble, and is gradually precipitated as a deep blue granular powder. This precipitation is facilitated by brisk agitation of the liquid, or by the addition of lime-water, or an alkaline solution. The supernatant liquor is then drawn off, and the sedimentary matter after being heated is thrown upon a calico filter, where it is washed. The indigo is then removed from the filter, pressed, and cut into cubical cakes, dried and sent into the market.

Indigo has been used as a remedial agent in epilepsy, and also in infantile convulsions, chorea, hysteria, and amenorrhœa." (Bentley and Trimen.)

IPECACUANHA, WILD OR BASTARD, See ASCLEPIAS CURASSAVICA.

IPOMŒA HEDERACEA, Jacq.

KALADANA.

Tropical and sub-tropical regions of both hemispheres. An annual herbaceous twining plant; leaves simple, alternate, cordate, 3-lobed; flowers usually bright pale blue, but sometimes purplish, pink, or white. *(Convolvulaceæ.)*

SEEDS. An effectual, quickly operating, safe cathartic. Dose of the powdered seeds is from 30 to 50 grains.

IPOMŒA PURGA, Hayne.

JALAP.

Native of country round Jalapa in West Mexico, about 6,000 feet elevation. Introduced, and now naturalised at Cinchona, Jamaica.

An herbaceous perennial twiner; leaves alternate, cordate, corolla salver-shaped, purplish-pink; roots tuberous. *(Convolvulaceæ.)*

ROOTS, dried, form commercial jalap. A certain, powerful, and speedy drastic purgative. Valuable in habitual constipation, and in febrile and inflammatory affections. A hydragogue in dropsies; a derivative purgative in head affections.

IRON SHRUB. See SAUVAGESIA ERECTA.

IRON WOOD. See LAPLACEA HAEMATOXYLON.

JACK FRUIT. See ARTOCARPUS INTEGRIFOLIA.

JALAP. See IPOMŒA PURGA.

JOB'S TEARS. See COIX LACHRYMA.

JOHN CROW BUSH. See BOCCONIA FRUTESCENS.

JUNIPERUS BARBADENSIS, Linn.

JUNIPER CEDAR, BERMUDA CEDAR.

Native of West Indies, Bahamas, Bermuda. A large tree, with small needle-leaves. *(Coniferae.)*

Wood. "This wood usually grows at a considerable elevation. It is one of the fir family and not abundant; it grows tolerably straight with many side branches, to a diameter of about 12 to 16 inches It is one of the most beautiful of our ornamental woods. Furniture, ceilings, door pannelings &c., made of this wood are unsurpassed for beauty. The wood has a pleasing odour." (*Harrison.*)

"Distributed over a clearly defined area of the Blue Mountains, above 4,000 feet and under 6,000 feet. It is rarely found larger than one and a half feet diameter in girth, with a height of 40 feet. It is now a rare tree, having been cut wherever accessible. Gives a handsome light wood, with beautiful graining, used for furniture and interior ornamental house works. Steps have been taken to grow it at lower elevations, and the young trees seem healthy and grow quickly. Later on they are apt to get stag-headed and twisted, but their planting deserves encouragement." (*Hooper.*)

KALADANA. See Ipomea hederacea.

KITTUL FIBRE PALM. See Caryota urens.

KOLA. See Cola acuminata.

KHUS-KHUS GRASS. See Andropogon muricatus.

LACE BARK. See Lagetta lintearia.

LAGENARIA VULGARIS, Sw.

Bottle Gourd.

Native of Jamaica. A climbing plant, with tendrils; leaves roundish; fruit of various shapes and sizes, some 6 feet long. (*Cucurbitaceæ.*)

Leaves.—Decoction said to be purgative.

Fruit.—Shell used for holding water, &c.

LAGETTA LINTEARIA, Lam.

Lace Bark Tree.

Native of Jamaica. A tree, 25 to 30 feet high; leaves broad-ovate, 3 to 5 inches long; flowers white. (*Thymeleaceæ.*)

Bark.—"The bark produces a beautiful fibre, very strong and well suited for the most delicate textile purposes: when carefully drawn out or stretched by the hands a pentagonal and hexagonal mesh is formed in every respect like lace, and many ornamental things are made from it." (Harrison.)

LAGUNCULARIA RACEMOSA, Gr.

White Mangrove.

Native of West Indies, tropical America and west Africa.

A small tree; leaves simple, opposite, flowers white; nut ½ inch; seed germinating in nut. (*Combretaceæ.*)

Bark used for tanning, useful to combine with the active divi-divi.

LANCE WOOD. See BOCAGEA LAURIFOLIA AND B. VIRGATA.

LAPLACEA HÆMATOXYLON, Camb.

BLOOD WOOD, IRON WOOD.

Native of Jamaica. A tree, leaves simple; flowers large, white; seeds with a long wing at the top. (*Ternstræmiaceæ.*)

WOOD.—"This is a very hard close grained wood of a deep red color, it is not used in building on account of its hardness. I think it would be a very useful wood for small articles, such as ornaments, knobs, buttons, &c., and is just the kind wood of wood now being enquired after by manufacturers. An excellent dye is extracted from it." (Harrison.)

"In the Blue Mountains at over 5,000 feet a very handsome, heavy, fine grained timber with much the same qualities as boxwood. Yields a dye from the rich red heart-wood." (Hooper.)

LAUDANUM, DUTCHMAN'S. See PASSIFLORA MURUCUJA.

LEAF OF LIFE. See BRYOPHYLLUM CALYCINUM.

LEMON. See CITRUS MEDICA, var. LIMONUM.

LEMON GRASS. See ANDROPOGON CITRATUS.

LEMON, WATER. See PASSIFLORA LAURIFOLIA.

LILAC. WEST INDIAN. See MELIA SEMPERVIRENS.

LIME. See CITRUS AURANTIUM, var. BERGAMIA.

LIQUORICE, WILD. See ABRUS PRECATORIUS.

LOCUST TREE. See HYMENÆA COURBARIL.

LOGWOOD. See HÆMATOXYLON CAMPECHIANUM.

LOTUS BERRY. See BYRSONIMA CORIACEA.

LUCUMA MAMMOSA, Gr.

MAMMEE SAPOTA.

Native of West Indies and tropical America.

A tree; leaves simple, 6 to 8 inches long; flowers white; berry about 6 inches long.

"Generally distributed, but nowhere very common. A small stout tree, with a diameter up to 2 feet and height of 30 feet, gives a first-class timber and adapted to many uses, especially house construction, both exterior and interior, furniture, &c., fi e specimens in Serge Island House, Blue Mountain Valley." (Hooper.)

LUFFA ACUTANGULA, Roxb.

STRAINER VINE, LUFFA.

Tropics. A climbing plant, with tendrils; leaves 5-lobed; flowers yellow; fruit size of a cucumber, drying up so as to leave nothing but a dense framework of fibres with the flat black seeds. (*Cucurbitaceæ.*)

FRUIT, deprived of the rind, used for rubbing the flesh in the bath, and for making ornaments.

LYCOPODIUM CLAVATUM, Linn.

CLUB-MOSS, STAGSHORN MOSS.

Native of temperate and colder regions of the whole world, and high elevations in the tropics. It grows in Jamaica above 4,500 ft.

This plant is a near relation of Selaginella, the two genera belonging to the order *Lycopodiaceæ.* It is not one of the true mosses, though related to them.

The minute spores are of economic value.

Lycopodium, in medicine, is used as a dusting powder to excoriated surfaces. In pharmacy, it is used for enveloping pills.

It is chiefly used by the pyrotechnist, and for producing artificial lightning at theatres. Thrown into flame, it produces an instantaneous flash.

The principal constituent is a fixed oil, to the extent of 47 per cent. It remains liquid at 5° F.

MACARY BITTER. See PICRAMNIA ANTIDESMA.

MACLURA TINCTORIA, Don.

FUSTIC.

Native of Jamaica, and tropical South America.

A tree with simple leaves, and minute flowers. (*Urticaceæ.*)

WOOD exported as a dyewood.

"A very tough close-grained and heavy wood of a burnt sienna colour, used for felloes of wheels." (Harrison.)

"On the coast, but on higher grounds generally than the logwood. A large tree with a diameter ranging up to 3 feet and a height of 40 feet. It yields a yellow brown timber, tough and close grained, used in carriage building, but largely exported as a dyewood. Average export for past 19 years 2,750 tons, the largest in one year 4,800 tons." (Hooper.)

MAHOE, BLUE. See HIBISCUS ELATUS.

MAHOE, CONGO. See HIBISCUS CLYPEATUS.

MAHOE, MOMNTAIN. See HIBISCUS ELATUS.

MAHOE, SEA-SIDE. See HIBISCUS TILIACEUS.

MAHOGANY. See SWIETENIA MAHAGONI.

MAIDEN PLUM. See COMOCLADIA INTEGRIFOLIA.

MAIZE. See ZEA MAYS.

MAJOE BITTER. See PICRAMNIA ANTIDESMA.

MAMMEA AMERICANA, Linn.

MAMMEE.

Native of West Indies and tropical America.

A spreading tree, 40 to 60 feet high; leaves simple, opposite; flowers large, white, fragrant; fruit larger than an orange, russet-brown. (*Guttiferæ.*)

WOOD "remarkably durable, well adapted for house building, posts and piles; stands damp. It is beautifully grained and used for fancy work." (Harrison.)

BARK. Gum applied to extract jiggers; "dissolved in lime-juice, it destroys maggots in sores at a single dressing; and an infusion of the bark is astringent, and is useful to strengthen the recent cicatrices of sores."

FLOWERS. "A liqueur has been obtained by distillation from the flowers infused in spirits of wine, known in Martinique by the name of "*Crême des Creoles.*" (Macfadyen.)

FRUIT of a sweetish, somewhat aromatic taste, and of a peculiar odour.

MAMMEE SAPOTA. See LUCUMA MAMMOSA.

MANDIOC. See MANIHOT UTILISSIMA.

MANGIFERA INDICA. Linn.

MANGO.

Native of the East Indies.

A tree; leaves, simple; flowers small.

WOOD: "Distributed over the cleared hills on the metamorphic soils; planted in hedge rows and around habitations. Introduced in eighteenth century from the East Indies, and now growing spontaneously in the interior of the island up to 4,000 feet; diameter up to four feet, height thirty feet. As a timber it has only a few special uses." (Hooper.)

MANGO. See MANGIFERA INDICA.

MANGROVE, RED. See RHIZOPHORA MANGLE.

MANGROVE, WHITE. See LAGUNCULARIA RACEMOSA.

MANIHOT UTILISSIMA, Pohl.

CASSAVA, MANIOC, MANDIOC.

Native probably of Brazil.

A half-shrubby perennial, with very large yellowish roots filled with a milky juice, generally poisonous; leaves large, very deeply divided into 3 to 7 segments: fruit with six narrow, thick wings. (*Euphorbiaceæ.*)

There are a number of varieties, according to colour of stem and division of leaves. There is also one with a non-poisonous juice in the root. But the plant generally known as "Sweet Cassava," is without wings on the fruit, and has a reddish root. (*Manihot Aipi*, Pohl.)

BITTER CASSAVA ROOT abounds in a milky poisonous juice, and does not become soft by boiling or roasting.

SWEET CASSAVA ROOT has a non-poisonous juice, has tough portions in the centre, but becomes quite soft by boiling, and is eaten like potatoes.

Cassava Meal is prepared from both kinds. The root is grated, by which the cells, containing the juice and starch-grains, are broken up. The grated material is placed under pressure, sometimes with water pouring through it. The pressure squeezes out all the juice; while a certain proportion of the starch-grains passes over with the liquor. The substance left under pressure consists chiefly of the cell-walls broken up, but also of some starch grains. This is Cassava Meal which is dried on hot plates, and made into Cassava cakes. The liquor which passes away under pressure, being the pure juice only, or the juice mixed with water, is allowed to stand for some time, when the starch settles to the bottom, and the liquor is poured off. The starch-grains, as seen under the microscope, are mullar-shaped. This is Cassava starch proper, as distinguised from Cassava meal.

Tapioca is prepared by heating moistened cassava starch on hot plates. This process alters the grains, which swell up, many bursting, and then they agglomerate in small irregular masses.

Cassareep is the juice of the bitter cassava root, concentrated by heat, which also dissipates the volatile poisonous principle. The same is further flavoured with aromatics. Boiled with peppers and fish or meat, it forms the West Indian "pepper-pot."

MANILA HEMP. See MUSA TEXTILIS.

MANIOC. See MANIHOT UTILSSIMA.

MELICOCCA BIJUGA, Linn.

GENIP.

Native of Trinidad and tropical South America.

A large tree, 40 to 50 feet high; leaves pinnate; flowers very numerous, small, fragrant; fruit green, size of a pigeon's egg. (*Sapindaceæ.*)

FRUIT.—Pulp edible, of a sweet sub-acid slightly astringent taste. Nuts in the Caraccas, are roasted and eaten, like chestnuts.

MILK WITHE. See FORSTERONIA FLORIBUNDA.

MILKWORT. See POLYGALA PANICULATA.

MOMORDICA BALSAMINA, Linn.

CERASEE.

Tropics. A climbing plant with tendrils; leaves deeply lobed; flowers yellow; fruit bright orange, bursting, showing the seeds with scarlet covering. (*Cucurbitaceæ.*)

"A decoction of the root is said to act as a de-obstruent, and to promote the secretions of the liver and kidneys.

The leaves are a favourite potherb in India, and have the reputation of promoting the lochial discharge when scanty."

Fruit used in East to cure wounds. It is cut open, infused in sweet oil, which is then exposed to sun for a few days, and put up for future use. Dropped on cotton, it is a good vulnerary. The pulp is purgative, and an infusion with addition of carbonate of soda removes discolourations of the skin. (Macfadyen.)

MOMORDICA CHARANTIA, Linn., is very much like the above, and is also known as *Cerasee.* The fruit is saffron-coloured or red, and the bract is below the middle of the flower-stalk, and quite entire. It is supposed to possess the same medicinal virtues; it is one of the ingredients of "pepper-pot".

MORINGA PTERYGOSPERMA, Gaertn.

HORSE RADISH TREE.

Tropical Asia and Africa.

A tree, 12 to 20 feet high; leaves thrice-pinnate; flowers whitish; pod nearly a foot long, three-cornered; seeds three-winged. (*Moringeæ.*)

ROOT is a substitute for horse-radish. It vesicates, and may be applied pounded as a rubefacient.

WOOD affords a blue dye.

SEEDS yield an oil.

MORONOBEA COCCINEA. See SYMPHONIA GLOBULIFERA.

MARANTA ARUNDINACEA, Linn.

ARROWROOT.

Native of West Indies and Central America. An herbaceous perennial, with a creeping rootstock; flowering-stem 5 or 6 feet high; leaves with long sheathing stalks, generally enveloping the stem; flowers yellowish-white. (*Zingiberaceæ.*)

ROOT-STOCKS, when about a year old, are beaten up in a mortar to a dulp. The pulp is well washed in water, and the fibrous portions thrown away. The milky-looking liquid is strained, and then allowed to stand. The starch settles, and when the water has been drained off, it is dried in the sun. The starch-grains are convex, more or less elliptical.

Arrowroot is very nutritious, and easily digested.

It is a valuable demulcent in bowel complaints and diseases of the urinary organs.

MAST WOOD. See CATALPA LONGISSIMA.

MAURITIUS HEMP. See FURCRŒA GIGANTEA.

MAY POLE. See AGAVE MORRISII.

MEXICAN POPPY. See ARGEMONE MEXICANA.

MELIA SEMPERVIRENS, Sw.

HOOP TREE, WEST INDIAN LILAC.

Tropics. A shrub or low tree; leaves twice-pinnate; flowers showy, blue mixed with purple and white; berry yellow. (*Meliaceæ.*)

ROOT: bark—a powerful anthelmintic.

STEM; bark bitter and astringent.

FRUIT: "pulp mixed with lard, forms an ointment for Tenia. It also yields a valuable bitter fixed oil, which is administered internally for worms, and employed externally to foul ulcers, and as a liniment in rheumatic and neuralgic affections." (Macfadyen.)

Dried berries—anthelmintic.

MUCUNA PRURIENS, DC.

COWITCH, COWHAGE.

Cosmopolitan in the tropics. A twining plant, belonging to the Pea Family *(Leguminosæ)*; leaves compound with 3 leaflets; flowers of a dark purple tinged pale yellowish-green, hanging in clusters on long pendulous stalks; pod, covered with a thick felt of pale-reddish stinging hairs, which turn brownish when dry. The word cowitch is a corruption of the Hindustani name cowhage.

ROOT—infusion used by natives in India as a remedy in cholera.

A decoction of root is said to be diuretic.

PODS—young and tender, are cooked and eaten in India.

A vinous tincture of the pods was formerly used for dropsy.

HAIRS from the pod form a mechanical anthelmintic. Administered in the form of an electuary with honey or treacle, they are used for the expulsion of the soft-bodied intestinal worms.

MUCUNA URENS, DC.

YELLOW-FLOWERED COWITCH, HORSE-EYE BEAN.

Native of West Indies and tropical South America.

A plant like the Purple-flowered Cowitch, but the pod is marked by transverse ribs, and is partially covered with stinging hairs.

SEEDS, known as Horse-eye Beans, are used for ornamental purposes.

MUSA PARADISIACA, Linn.

PLANTAIN.

Tropics. The stem is green, the bracts purple inside: fruit requires cooking.

STEM formed of the leaf-stalks contains a fibre which might prove valuable as paper-stock. See Bulletin, XVII., 12, 13.

FRUIT an excellent vegetable, also yields plantain meal.

MUSA PARADASIACA, Linn., var. SAPIENTUM.

BANANA.

Tropics. The stem is purple-spotted, and the bracts green inside.

STEM yields fibre as in plantain.

FRUIT eaten uncooked, dried, made up into puddings, &c.; also yields a meal.

MUSA TEXTILIS, Luis Nee.

MANILA HEMP, QUILOT MANILA.

Native of Philippine Islands. This plant, the Abaca of the Phillippines, is very much like the Banana and Plantain. but the fruit is not edible. It is in cultivation in Castleton Botanic Gardens.

"The Abaca is cut when about one year and a half old, just before its flowering or fructification is likely to appear, as afterwards the fibres are said to be weaker. If cut earlier, the fibres are said to be shorter and finer. It is cut near its roots, and the leaves cut off just below their expansion. It is then slit open longitudinally, and the central peduncle separated from the sheathing layers of fibres, which are in fact the petioles of the leaves. Of these layers the outer are harder and stronger, and form the kind of fibre called *bandala* which is employed in the fabrication of cordage. The inner layers consist of finer fibres and yield what is called *lupis* and used for weaving the *nipis* and other more delicate fibres; while the intermediate layers are converted into what is called *tupoz*, of which are made web-cloths and gauzes, four yards long, of different degrees of fineness. These are universally used as clothing; some being so fine that a garment may be enclosed in the hollow of the hand." (Royle.)

MUSK OKRA. See HIBISCUS ABELMOSCHUS.

MUSK WOOD. See GUAREA SWARTZII.

MUSTARD, BASTARD. See CLEOME PENTAPHYLLA.

MYRISTICA FRAGRANS, Houtt.

NUTMEG.

Native of Eastern Moluccas.

A tree, 20 to 40 feet high, with simple alternate leaves: small pale yellow flowers, male and female on separate trees; fruit fleshy, yellow, bursting into two valves, and showing the dark-brown shell of the seed covered with the scarlet *mace*; the shell is brittle and contains the *nutmeg*. (*Myristicaceæ*.)

SEED—the Kernel or "nutmeg" and the "mace" are principally used for flavouring.

In medicine, nutmeg possesses aromatic, stimulant, and carminative properties; but in large doses it is narcotic.

Volatile oil of nutmeg obtained by distillation, may be used for same purposes as nutmegs.

Expressed oil applied in chronic rheumatism, paralysis, and sprains.

NASEBERRY. See ACHRAS SAPOTA.

NASEBERRY BULLET TREE. See SAPOTA SIDEROXYLON.

NECTANDRA EXALTATA, Gr.

TIMBER SWEET WOOD.

Native of Jamaica and Dominica.

A high tree; leaves simple, 3 to 4 inches long; flowers whitish, berry with a cupule half as long as itself. (*Laurineæ*.)

WOOD. "Common up to a height of 3,000 feet above the sea. A tree with a diameter up to two feet and height of 40 feet. It is used in coarse carpentry and coopering, but has few qualities to recommend it." (Hooper.)

NECTANDRA LEUCANTHA, Nees.

TIMBER SWEET WOOD, WHITE SWEET WOOD, LONG-LEAVED SWEET WOOD, SHINGLE WOOD.

Native of West Indies and tropical South America.

A tree; leaves simple, 4 to 9 inches long; flower whitish; berry dark blue. (*Laurineæ*.)

WOOD. "This tree grows straight to about 2 or 3 feet in diameter. It splits and saws freely and makes very good boards, but not very lasting. Shingles and staves are frequently made from it, as also from the common sweet wood." (Harrison.)

NEGRESSEE. See BUCIDA CAPITATA.

NEW ZEALAND FLAX. See PHORMIUM TENAX.

NICKER SEEDS. See CÆSALPINIA BODUCELLA, AND C. BONDUC.

NICOTIANA TABACUM, Linn.

TOBACCO.

Native of some part of Central or South America. An annual plant, with large leaves covered with viscid hairs and some glands, and dull, pink flowers. *(Solanaceæ.)*

LEAVES cured form tobacco.

As a medicine, tobacco owes its value to its powerfully sedative and antispasmodic properties.

NO EYE PEA. See CAJANUS INDICUS.

NUT GRASS. See CYPERUS ROTUNDUS..

NUTMEG. See MYRISTICA FRAGRANS.

OAK, FRENCH. See CATALPA LONGISSIMA.

OCHRA. See HIBISCUS ESCULENTUS.

OCHRA, MUSK. See HIBISCUS ABELMOSCHUS.

OCHROMA LAGOPUS, Sw.

DOWN TREE, CORK WOOD.

Native of West Indies, and tropical America. A tree, 20 to 40 feet high, leaves simple, flowers large pale-reddish or yellowish-white; seeds enveloped in wool.

BARK used for making ropes.

WOOD soft, only fit for use as a substitute for cork.

SEEDS—The down used for stuffing pillows, &c. Possibly it might be made into cloth, &c.

OIL PLANT. See RICINUS COMMUNIS.

OLIVE BARK TREE. See BUCIDA BUCERAS.

OLIVE, WILD. See BUCIDA BUCERAS, and B. CAPITATA

ONIONS. See ALLIUM CEPA.

OPUNTIA COCCINELLIFERA, Mill.

COCHINEAL CACTUS.

Native of Jamaica and Mexico.

A cactus, 6 feet and more in height, or, generally without spines; flowers crimson. *(Cacteæ.)*

On this plant, and on *O. Tuna*, the cochineal insect lives from which carmine is prepared. The insect does not thrive in Jamaica.

OPUNTIA TUNA, Mill.

PRICKLY PEAR.

Native of West Indies, and tropical America.

A cactus 3 or 4 feet high, spines 4 or 5 in a cluster, yellowish flowers yellow of 3 or 4 inches diam.

"The plant roasted is applied as a poultice to indolent swellings and foul sores.

The young joints are sometimes employed as a substitute for ochras to thicken soup.

The fruit is insipid, but is said to possess astringent properties. It is principally employed to give a crimson colour to liqueurs, and to the fruits used in confectionery." (Macfayden.)

Plants used for hedges.

On it, and on *O. coccinellifera,* the coccus insect lives, which yields the cochineal dye.

ORANGE. See CITRUS AURANTIUM.

ORYZA SATIVA, Linn.

RICE.

Native of India and China. An annual grass. (*Gramineæ.*)

SEEDS (Rice) contain a larger proportion of starch (85 to 90 per cent.) than other cereals, much less nitrogenous substances (7 per cent.), and less of fatty matters (0.8 per cent.), and inorganice constituents. Rice is therefore less nutritive, and yet it is more largely used as food than any other grain.

PAPAW. See CARICA PAPAYA.

PARROT WEED. See BOCCONIA FRUTESCENS.

PASSIFLORA EDULIS, Sims.

MOUNTAIN SWEET CUP.

Native of Andes. Naturalised in Blue mountains.

A passion-flower with lobed leaves and whitish flowers; fruit egg-shaped, purplish, with hard shell. (*Passifloraceæ*)

FRUIT—edible pulp.

PASSIFLORA LAURIFOLIA, Linn.

WATER-LEMON, POMME D'OR.

Native of West Indies and tropical America.

A passion-flower; leaves oval, entire; flowers white with red blotches crown violet with white streaks; fruit egg-shaped, of an orange-yellow colour, and soft rind. (*Passifloraceæ.*)

FRUIT—edible pulp.

PASSIFLORA MALIFORMIS, Linn.

SWEET CUP.

Native of West Indies and tropical America.

A passion-flower; leaves ovate, entire; flowers variegated; fruit globular, green with yellow tint, very hard shell. *(Passifloraceæ.)*

FRUIT—edible pulp.

PASSIFLORA MURUCUJA, Linn.

DUTCHMAN'S LAUDANUM.

Native of Jamaica, Cuba, Hayti.

A passion-flower; leaves 2-lobed; flowers crimson, crown entire; fruit size of an olive, of a fleshy colour. (*Passifloraceæ.*)

"The flowers are said to possess narcotic properties, and a syrup made with them, and also a tincture, have been employed as substitutes for the syrup of poppies and for laudanum." (Macfadyen.)

PASSIFLORA QUADRANGULARIS, Linn.

GRANADILLA.

Native of West Indies and tropical America.

A passion-flower with four-winged stem; leaves ovate-roundish; flowers large, petals rosy, crown violet; fruit five inches long, greenish-yellow, with soft thick rind. *(Passifloraceæ.)*

ROOT, large and fleshy, edible like yam.

FURIT, pulp edible, rind edible when cooked.

PAULLINIA CURASSAVICA, Jacq.

SUPPLE JACK.

Native of Jamaica and New Granada.

A climbing shrub, a few feet high; leaves compound, ternate; flowers small, white, in clusters; seed-vessel, ½ inch long, red; seeds, black, half-enclosed in a brownish-white covering.

BRANCHES used as riding switches.

SEEDS "possess the property of intoxicating fish." (Macfadyen.)

PEA FLOWER. See CLITORIA TERNATEA.

PELTOPHORUM LINNÆI. Benth.

BRAZILETTO.

Native of Jamaica.

A low tree; leaves twice-pinnate; small, yellow flowers; pod, flat, 3 inches long. (*Leguminosæ.*)

WOOD.—"This wood is much used for spokes of wheels and in carriage building generally; being of a bright red colour and taking a high polish. It is also much used for ornamental cabinet work." (Harrison.)

"In most parts of the Island, notably in the district of Vere, Clarendon. Trees up to 2½ feet in diameter; tall, straight growth; gives a handsome ornamental timber, useful in carriage building and cabinet work. This has been exported in small quantities." (Hooper.)

"It is full of resin, and gives, on infusion, a fine full tincture." (Macfadyen.)

PHASEOLUS LUNATUS, Linn.

SUGAR BEAN, BROAD BEAN, HIBBERT BEAN, LIMA BEAN.

Native of West Indies and tropical America. A twining plant; leaves compound with one to three leaflets; flowers usually greenish-white; pod scimitar-shaped, compressed; seeds purple or white. (*Leguminosæ.*)

Beans—of an excellent quality.

PHASEOLUS VULGARIS, Linn.

KIDNEY BEAN, HIBBERT PEA, HARICOT BEAN, FRENCH BEAN.

Native of tropical Asia. A twining plant; leaves compound, with 1 to 3 leaflets; flowers usually white; pod straightish; seeds variable in colour. (*Leguminosæ.*)

BEANS—edible. "The Kidney Bean is nearly twice as nutritious as wheat." (Mueller.)

PHORMIUM TENAX, Forst.

NEW ZEALAND FLAX.

Native of New Zealand, Norfolk Is., Chatham Is., and Auckland Is. The Flax Lily of New Zealand has long narrow leaves, 3 to 6 feet long. The branched flower-spike rises from 6 to 16 feet, and the flowers are of an orange colour. (*Liliaceæ.*)

It is under cultivation in the Hill Garden, Cinchona, where it flowered in 1889. Erperiments are now being made as to its adaptability for the plains. It grows on inferior ground, but thrives best on rich soil.

"A strong decoction of the root and leaf-bases is used in surgery for dressing wounds with a view of producing ready and healthy granulation."—(F. A. Monckton.)

The leaves give a very large percentage of fibre, viz., 15.20 percent., compared with the 3 to 5 per cent. in the Agaves. The gummy matter, however, requires that the fibre should be treated with some such substance as sulphite of soda in order to make it of superior quality. The fibre is naturally white, soft, and of a silky lustre. It is used for making ropes, and the refuse is an excellent paper material.

PICRÆNA EXCELSA, Lindl.

BITTER ASH, BITTER WOOD, JAMAICA QUASSIA.

Native of West Indies. A tree up to 60 feet, in diameter up to 2 feet; with pinnate leaves, small flowers of a pale yellowish-green, and black berries. (*Simarubaceæ.*)

Wood is the Quassia of commerce in form of chips. It is a pure tonic and stomachic. "It is a valuable remedy in atonic dyspepsia, in debility, and in convalescence after acute diseases. It has also been administered with success as an anti-periodic in fevers; and as an enema to destroy thread-worms in children As it contains no tannic acid it is frequently given in combination with chalybeates." (Bentley & Trimen.)

"Bitter-cups" are made from the wood.

An infusion used to preserve animal matters from decay; and paper dipped in an infusion sweetened with sugar, is used to destroy flies.

Quassia used by brewers as a substitute for hops.

PICRAMNIA ANTIDESMA, Sw.

MAJOE BITTER, MACARY BITTER.

Native of West Indies, and tropical America.

A shrub about 8 feet high; leaves pinnate; flowers very small, whitish-green; berries, first scarlet, then black in hanging bunches. (*Rutaceæ.*)

BARK—a bitter; "an alterative in constitutional affections, connected with syphilis and yaws, and as a tonic in debility of the digestive organs, and in intermittent fever." (Macfadyen.)

PIGEON WOOD. See DIOSPYROS TETRASPERMA.

PIMENTA ACRIS, Wight.

WILD CINNAMON, WILD CLOVE, BAYBERRY.

Native of West Indies and Venezuela. A tree 30 to 40 feet high; leaves simple, opposite, 2 to 3½ inches long, covered with minute dots beneath; flowers small, whitish with a tinge of red; berry blackish. (*Myrtaceæ.*)

LEAVES yield "oil of bay," which is one of the ingredients of Bay-rum. This preparation is official in the United States Pharmacopœia, and is said to be the spirit prepared by distilling rum with the leaves. Three-fourths of the Bay-rum used in the States is not imported, but made up from the oil distilled from the leaves. The tree is cultivated in India for the sake of its aromatic leaves and berries. (Hooker.)

Bay-rum is a refreshing perfume in nervous affections, and to sprinke in sick rooms. It is also used for perfumery.

PIMENTA OFFICINALIS, Lindl.

PIMENTO, ALLSPICE.

Native of West Indies and tropical America. A tree, 30 feet high, with a very smooth, light grey bark; leaves simple, opposite, 4-6 inches long, with minute dots beneath; flowers small, white. (*Myrtaceæ.*)

STEMS.—Young, are made into walking-sticks and umbrella handles. "A valuable close-grained wood, Male trees chiefly cut down, the others yielding the well-known Jamaica allspice. Diameter of the tree up to one foot." (Hooper.)

Berries, collected of full size, though unripe, and dried in sun for export. They yield, by distillation with water, a volatile oil, which is aromatic, carminative, and stimulant, promoting digestion.

Pimento is used as a spice, and the ripe berries in the preparation of a liqueur—" Pimento Dram." The oil is used also for perfumery.

Leaves can be used for tanning.

PINDAR. See Arachis hypogæa.

PINE APPLE. See Ananas sativa.

PISCIDIA ERYTHRINA, Linn.

Dogwood.

Native of West Indies, and tropical America. A tree, 15 to 30 feet high; with pinnate leaves; flowers whitish with a purplish tinge; pod with longitudinal wings. (*Leguminosæ.*)

Bark of the root used medicinally. It is an intense narcotic, and relieves toothache, when placed in the hollow of carious teeth. The tincture is often used in the United States instead of opium.

Bark of the stem, used to intoxicate fish, by pounding it up, and throwing it into the deep part of the stream. A decoction of the bark cures the mange in dogs. (Barham)

Wood—"A very useful wood, tough and elastic, much used for felloes of wheels and for cart and carriage frames and other work requiring a tough wood." (Harrison.)

"Found with the Yoke wood, and stretches up to a greater altitude. A small tree growing to a diameter of two feet and a height of 40 feet. A valuable wood, tough and elastic, used in cart building for bodies and wheels. The root bark used in the United States as a narcotic." (Hooper.)

PITHECOLOBIUM FILICIFOLIUM, Benth.

Wild Tamarind.

Native of West Indies and Central America.

A lofty tree; leaves twice-pinnate; flowers whitish in long-stalked heads; pod twisted scarlet, blood-colored within, with black seeds. (*Leguminosæ.*)

Wood.—" An excellent timber wood, much used in building, found all over the Island, grows straight up to 3 feet in diameter, saws freely, not too hard for general work, takes a fine polish and makes pretty flooring and ceiling." (Harrison.)

" Generally distributed over the Island, except on the high levels of the Blue Mountains. Occasionally this species attains considerable size, up to 12 feet in girth (Good Intent, Manchester) and 60 feet of stem, but in general it scarcely exceeds a diameter of three feet. It is in general use for floorings and ornamental work." (Hooper.)

PITHECOLOBIUM SAMAN, Benth.

Guango.

Native of tropical America. Naturalised in Jamaica.

A large spreading tree, often 6 feet in diameter, with compound leaves, reddish flowers, and long pods. (*Leguminosæ.*)

Wood hard and ornamental, but it is cross-grained and difficult to saw.

Pods are a very fattening fodder. The trees form a good shade in pastures.

"Found on the coast, but growing to perfection on the alluvial plains of St. Catherine and Clarendon, and in the Plantain Garden Valley. A large spreading tree, with a short stem and a diameter up to five feet at the base. Grown in pastures and yields a very sweet pod eagerly eaten by horses and stock. Introduced with cattle from the mainland." (Hooper.)

PLANTAIN. See Musa paradisiaca.

PLUM, DATE. See Diospyros tetrasperma.

PLUM, MAIDEN. See Comocladia integrifolia.

PODOCARPUS CORIACEUS, Rich.

Yacca.

Native of West Indies. A tree 50 feet high; leaves lanceolate, with acute tip. (*Coniferæ.*)

Wood. "This is one of our most prized ornamental woods and much used in furniture and cabinet work. The most flowery specimens grow on the Blue Mountains, where it is crooked and cross-grained. In other elevated parts of the island it is straight and yields capital building and furniture wood, but not so well adapted for ornamental purposes as that grown on the Blue Mountains. Its largest growth is about 18 inches in diameter." (Harrison.)

"Blue Mountains, elevations above 4,800 feet to the summit. Up to two feet in diameter in favourable localities, with stem of 40 feet. It is well distributed over the limited area it occupies. All accessible large trees have been cut. The timber gives beautifully-marked planks, used in cabinet work, also in the interior of finishing of dwelling-houses." (Hooper.)

PODOCARPUS PURDIEANUS Hook.

Yacca.

Native of Jamaica and Cuba.

A large tree, 120 feet high; leaves narrow, with blunt tip. (*Coniferæ.*)

WOOD: "The Central range on Mount Diabolo. Also reported to be in St. Ann's and Manchester. I have not seen identified specimens of this timber. It is reported of different graining to the last named, straighter and probably larger than its congener. Not much timber available." (Hooper.)

POISON WOOD. See RHUS METOPIUM.

POLYGALA PANICULATA, Linn.

MILKROOT.

Native of W. Indies, Central and S. America. An annual herb, with small leaves and purplish or white flowers. *(Polygaleæ.)*

Properties—sudorific and diuretic,—either as decoction or infusion.

POMEGRANATE. See PUNICA GRANATUM.

POMME D'OR. See PASSIFLORA LAURIFOLIA.

POOR MAN'S WEATHER GLASS. See ANAGALLIS ARVENSIS.

POPPY, MEXICAN. See ARGEMONE MEXICANA.

PORTULACA OLERACEA, Linn.

COMMON PURSLANE.

Temperate and tropic zones. An annual weed, with fleshy leaves, and yellow flowers. *(Portulaceæ.)*

"As a vegetable it has the reputation of being cooling, antiscorbutic, diuretic, &c., and peculiarly adapted for warm weather." (Macfadyen.)

PORTULACA PILOSA, Linn.

PURPLE-FLOWERED PURSLANE.

Native of West Indies and tropical America. An annual weed with fleshy leaves, and purple flowers. *(Portulaceæ.)*

"The leaves are intensely bitter to the taste, and have been used as a diuretic and stomachic, as well as an emmenagogue." (Macfadyen.)

PRICKLY PEAR. See OPUNTIA TUNA.

PRICKLY YELLOW. See ZANTHOXYLUM CLAVA-HERCULIS.

PROSOPIS JULIFLORA, DC.

CASHAW.

Native of Jamaica and tropical America.

A tree, 30-40 feet high, spiny; leaves twice-pinnate; flowers yellowish, delicately fragrant; pod with sweetish pulp.

STEM:—a gum like gum arabic, obtained in considerable quantities by wounding stem and large branches.

"A very hard wood, splits readily, makes everlasting shingles; it is the most abundant wood on the plains of the south side and is the chief firewood. It is much used for fencing and the sleepers of the original line of the Jamaica Railway are of this wood. The branches being very crooked they are well adapted for the knees of boats and ship building generally." (Harrison.) It is excellent for fuel.

Stock of every kind feed on pods, leaves, and shoots. In dry weather the pods are said to be as nutritious as corn, but after rains, horses often die, apparently, from the germination of the seed in the stomach; remedies,—alkaline solution to absorb carbonic acid, and saline purgatives. (Macfadyen.)

"Found in the alluvial deposits along the south coast. A small tree up to 20 feet in height, with a diameter of from 8 to 12 inches. Is a very desirable strong timber, the uses to which it can be put being only limited by the small size it attains. It makes good posts for telegraph work, sleepers for railway lines and knees in boats and ships." (Hooper.)

PRUNE TREE. See PRUNUS OCCIDENTALIS.

PRUNUS OCCIDENTALIS, Sw.

PRUNE TREE.

Native of West Indies, Guatemala, and Panama.

A high tree, with simple leaves, 4 to 6 inches long; flowers yellowish-white; fruit like a plum, but very little flesh. (*Rosaceæ.*)

WOOD, "of a red colour, resembling cedar, and is very hard and durable; from its taking a fine polish, it makes a beautiful flooring for houses." (Macfadyen.) "A hard and durable wood, attains a size of from 2 to 3 feet in diameter, lasts well in water, is therefore good for piles; it is considered an excellent timber for building. (Harrison.") "Found in the higher level of the hills. Up to three feet in diameter. Yields a good building timber, and lasts under water." (Hooper.)

"It is from the kernels that the celebrated liqueur, the *Noyau* of Martinique, is prepared. They yield a flavour much superior to that of the peach, being rich, oily, and nutty, combined with that of prussic acid." (Macfadyen.)

BARK, "has an astringent taste, with a strong flavour of prussic acid, and is used in manufacturing an inferior description of *Noyau*, known by the name of Prune-dram." (Macfadyen.)

PSIDIAM GUAVA, Radd.

GUAVA.

Native of West Indies and tropical America.

A low tree; leaves simple; flowers white, large. (*Myrtaceæ.*)

BARK can be used for tanning. A decoction of bark, young leaves, and fruit,—astringent, useful in diarrhœa and dysentery.

FRUIT—Stewed, and made into jelly.

PSIDIUM MONTANUM, Sw.

Mountain Guava.

Native of Jamaica.

A lofty tree, sometimes 100 feet in height; bark very smooth, ash-coloured; and leaves simple, 3 inches long; flowers large, white,—with the odour of bitter almonds; berry sour, size of a cherry. (*Myrtaceæ.*)

Wood "is highly esteemed, affording a timber of the hardest description, with the grain beautifully variegated." (Macfadyen.)

"A hard white wood, grows straight to about 2 feet in diameter, not much used in building, perhaps on account of its hardness and cross grain, and because when used as posts, it rots quickly in the earth." (Harrison.)

"At elevations from 3,000 to 6,000 feet, not a very common timber tree and useful only for gun-stocks and such articles as require hard tough wood, grows up to two feet in diameter." (*Hooper.*)

PUNICA GRANATUM, Linn.

Pomegranate.

Native of Persia, Cabul, N. W. India.

A bush or small tree, with simple leaves, crimson flowers, and reddish-yellow fruit. (*Lythraceæ.*)

Bark of root—a vermifuge; also used in chronic diarrhœa, and dysentery.

Flowers, called "Balaustine flowers," possess astringent properties.

Fruit used as a dessert fruit, and may be eaten "as a slightly astringent and refreshing refrigerant in some febrile affections, especially those of the bilious type."

Rind of fruit—astringent, and may be used externally and internally; highly esteemed in India for diarrhœa and chronic dysentery, usually combined with opium.

PURSLANE. See Portulaca.

PURSLANE, SEA-SIDE. See Sesuvium portulacastrum.

QUASSIA, JAMAICA. See Picræna excelsa.

QUILOT. See Musa textilis.

RAMIE. See Bœhmeria nivea.

RAMOON. See Trophis americana.

RAPHIA RUFFIA, Mart.

RAFFIA—FIBRE PALM.

Native of Madagascar.

The Raphia palm grows in brackish swamps The trunk is not large, but the pinnate leaves are often 50 feet in length.

There are young palms in the Castleton Botanic Gardens.

FIBRE yielded by the decaying spathes of the leaves.

RED HEAD. See ASCLEPIAS CURASSAVICA.

REDWOOD. See ERYTHROXYLUM AREOLATUM.

RHIZOPHORA MANGLE, Linn.

RED MANGROVE.

Native of West Indies, tropical America and West Africa. A tree, 20 to 40 high, forming thickets along swampy shores by sending down roots from their branches and from the seeds while still in the fruit. (*Rhizophoreæ.*)

WOOD "makes a good post, and lasts well as piles." (Harrison.)

BARK used for tanning, especially sole leather.

RHUS METOPIUM, Linn.

JAMAICA SUMACH, POISON WOOD.

Native of Jamaica and Cuba. A tree, 15 to 40 feet high; leaves compound, with three leaflets, or only one; flowers small; berry scarlet. (*Anacardiaceæ.*)

"Common on limestone hills. Affords a good timber of light colour and texture." (Hooper.)

RICE. See ORYZA SATIVA.

RICINUS COMMUNIS, Linn.

CASTOR OIL. OIL PLANT.

Native probably of Africa; found throughout the tropics.

A bush or small tree; leaves cut into 7 to 11 segments; flowers monœcious on a long stalk, male flowers with numerous stamens, female with carmine style; fruit spiny splitting into six valves. (*Euphorbiaceæ.*)

SEEDS yield the oil.

The method of preparing it in the West Indies by boiling, skimming off and straining the oil, produces an inferior article. "All the oil now consumed in England, India, and the United States, is obtained by expression. All processes in which a high temperature is employed are considered objectionable from increasing the acridity of the oil. In India, in order to extract the oil, the seeds are at first gently crushed between rollers, and after the seed-coats or husks, and unsound seeds have been removed by hand-picking, the clean kernels are submitted to pressure in an hydraulic press; and the oil thus obtained is first heated with water until the water boils, by which the albuminous matters are separated as a scum; and the oil is then finally strained through flannel." (Bentley and Trimen.)

Besides its general use as a purgative, it is recommended by Dr. Johnson as an eliminant in malignant cholera.

LEAVES, in the form of decoction or poultice, are used as a lactagogue.

ROSE WOOD. See AMYRIS BALSAMIFERA.

RUBUS ALPINUS. Macf.

JAMAICA ALPINE BLACKBERRY.

Native of Jamaica.

A bramble, with 3 leaflets to each leaf, not hairy; branches with a purplish tinge. *(Rosaceæ.)*

FRUIT, small.

RUBUS JAMAICENSIS, Sw.

JAMAICA BLACKBERRY.

Native of Jamaica.

A bramble, with 3-5 leaflets to each leaf, covered with silky hairs beneath. *(Rosaceæ.)*

FRUIT, palatable. "Infused in spirit with the bruised kernels of the Prune Tree, and sweetened with sugar, a liqueur is obtained, not inferior to, and not to be distinguished from, the Copenhagen Cherry Brandy." (Macfadyen.)

SACCHARUM OFFICINARUM, Linn.

SUGAR CANE.

Native of India and China. A large perennial grass, with a thick root-stock, and numerous stems. *(Gramineæ.)*

Preparation of Raw Sugar:—"The ripe canes are cut down to the ground, stripped of their leaves, and subjected to pressure between two rollers, or in some other suitable way. The cane juice thus obtained, is clarified by the combined use of lime and heat. The heat coagulates any albumen which may be present; and the lime neutralises the free acid, and combines with a peculiar albuminous body not coagulable by heat or acids, and forms with it a coagulum, the separation of which is promoted by the heat. Part of it rises to the top as a scum, and the remainder subsides. The clarified juice is then drawn off into the boiler, evaporated and skimmed. When it has acquired a proper tenacity and granular aspect, it is emptied into a cooler and allowed to crystallize or *grain*. The concrete sugar is then placed in casks perforated with holes in the bottom; and the sugar is left to drain for 3 or 4 weeks. It is then packed in hogsheads, forming *raw* or *muscovado* sugar. The drainings or uncrystallized portion of sugar constitute *molasses*.

Sugar is nutritious, but in consequence of not containing nitrogen, it is not capable in itself of supporting life. It is a powerful antiseptic, and is largely used for preserving meat and fruit." (Bentley & Trimen.) Molasses is capable of fermentation, and then by distillation yields *rum*.

SANDERS, GREY MOUNTAIN. See Bucida sp.

SANDERS, YELLOW. See Bucida capitata.

SANTA MARIA. See Calophyllum Calaba.

SAPINDUS SAPONARIA, Linn.

Soap Berry.

Native of Jamaica and Venezuela. A tree, 15 to 30 feet high: leaves pinnate, with leaf-stalk broadly winged between the leaflets; flowers small, white; berry size of a cherry; seed black.

Seeds. The fleshy covering, and in a less degree the root, make a lather in water, and serve all the purposes of soap, but are apt to injure the cloth. The seeds were formerly imported into England for waistcoat buttons; and are often strung as beads, &c.

They appear to be possessed of medicinal properties. Bruised and thrown into water, they kill fish. Given to fowl, they are said to be a preventive against "fowl-yaws." "A tincture, prepared by infusing the bruised berries in spirits, is often used as an embrocation in rheumatism;" and is also said to be of great value in chlorosis. (Macfadyen.)

SAPODILLA. See Achras Sapota.

SAPOTA, MAMMEE. See Lucuma mammosa.

SAPOTA SIDEROXYLON, Gr.

Naseberry Bullet-tree.

Native of Jamaica.

A tall tree, like the naseberry, but leaves twice as large and fruit only ½ inch in diameter. *(Sapotaceæ.)*

Wood: "This tree grows much straighter and taller than the bullet-tree, and is considered a better timber. It is very heavy and wears well under water. It grows to 3 or 4 feet in diameter." (Harrison.) "Generally in the central and west-central parts of the island. A large tree, up to four feet diameter and 60 feet of stem, but occasionally larger (at Peru, St. Elizabeth, six feet diameter, grown isolated in a pasture). A good wood for general purposes, but very heavy." (Hooper.)

SARSAPARILLA. See Smilax officinalis.

SAUVAGESIA ERECTA, Linn.

IRON SHRUB.

Native of West Indies, Central and S. America.

An annual herb, 6 to 9 inches high, with small white flowers.

"It is very mucilaginous, and the infusion has been used in Brazil for complaints of the eye, in Peru for dysentery, and, in some of the West Indian Islands for irritability of the bladder." (Macfadyen.)

SCREW TREE, JAMAICA. See HELICTERES JAMAICENSIS.

SECHIUM EDULE, Sw.

CHOCHO.

Cultivated in the West Indies.

A climbing plant, with tendrils; roots large and fleshy; leaves large, simple; flowers yellow; fruit 4 or 5 inches long.

ROOT can be used like the yam.

FRUIT—a most wholesome vegetable. It is also made into tarts with addition of lime.

SESUVIUM PORTULACASTRUM, Linn.

SEA-SIDE PURSLANE.

Tropics. A perennial weed, with fleshy leaves; flowers with no petals, but the sepals are coloured purple within. *(Ficoideæ.)*

Pickled, it is like samphire.

A decoction is useful as a gargle.

An alkali might be obtained from it.

SIDA CARPINIFOLIA, Linn.

BROOM WEED.

Tropics. A somewhat shrubby plant, about a foot high, with simple leaves and shortly-stalked yellow flowers. *(Malvaceæ.)*

Branches tied together, used as a broom.

"The leaves and tender shoots are made use of as a substitute for soap. Rubbed up with water, they form a lather, which may be employed in shaving, when the skin is in an irritable state, not admitting of the use of soap." (Macfadyen.)

SIDA RHOMBIFOLIA, Linn.

BROOM WEED.

Tropics. A somewhat shrubby plant, from one to three feet high, with simple leaves, and pale yellow flowers with stalks nearly an inch long. *(Malvaceæ.)*

Branches used as a broom.

A good fibre can be extracted from the stems.

SILK COTTON TREE. SEE ERIODENDRON ANFRACTUOSUM.

SILK GRASS. SEE FURCRŒA.

SIMARUBA GLAUCA, D.C.

BITTER WOOD, BITTER DAN.

Native of West Indies, and Florida.

A tree about 20 feet high, with pinnate leaves, small yellow flowers, and purple berries. (*Simarubaceæ.)*

BARK of root is a tonic, used in dysentery, after fever has abated, in diarrhœa, and dyspepsia.

SISAL HEMP. SEE AGAVE RIGIDA.

SLOANEA JAMAICENSIS.

GREENHEART.

Native of Jamaica.

A tree 60 to 100 feet high, leaves simple; flowers without petals, seed enveloped in scarlet pulp.

"In the interior limestone hills of Manchester and St. Elizabeth. A tall tree, with diameter rarely exceeding 18 inches. A heavy timber, with dense deep-coloured heart-wood known as an ebony." (Hooper.)

SMILAX OFFICINALIS, Kunth.

SARSAPARILLA.

Native of tropical America. A large perennial climber, with large rootstock; which gives off numerous roots 6 to 8 feet long; leaves a foot long, cordate; flowers small and inconspicuous.

ROOTS are the official part of the plant. Sarsaparilla is alterative, tonic, diaphoretic, and diuretic. Employed in syphilitic affections, in chronic forms of rheumatism, gout, scrofulous affections, and skin diseases.

"Jamaica Sarsaparilla" is alone official in the pharmacopœias of Britain and India. The name, however, does not mean that it is grown in Jamaica, but only that it was exported from Central America by way of Jamaica. The plant is cultivated to a small extent in the parish of St. Elizabeth.

SOAP BERRY. SEE SAPINDUS SAPONARIA.

SOAP WOOD. SEE CLETHRA TINIFOLIA.

SORREL, RED. SEE HIBISCUS SABDARIFFA.

SOUR SOP. SEE ANONA MURICATA.

SPANISH ELM. SEE CORDIA GERASCANTHOIDES.

SPIGELIA ANTHELMIA, Linn.

WORM GRASS.

Native of W. Indies and tropical S. America.

An annual herb, with opposite simple leaves, and small purplish flowers. *Loganiaceæ.)*

The root, and whole plant,—anthelmintic; particularly efficacious against *Lumbrici* (Round Worms).

STAG'S HORN MOSS. See LYCOPODIUM CLAVATUM.

STIPA TENACISSIMA, Linn.

ESPARTO GRASS.

Native of the shores of the Mediterranean. A rush-like grass, growing in sandy districts. *(Gramineæ.)*

The plant has been used from remote times for making baskets, hats, ropes, &c., and of late years an immense trade has sprung up in consequence of its utilisation as paper stock.

STRAINER VINE. See LUFFA ACUTANGULA.

STRAMONIUM. See DATURA STRAMONIUM.

SUGAR CANE. See SACCHARUM OFFICINARUM.

SUMACH, JAMAICA. See RHUS METOPIUM.

SUPPLE JACK. See PAULLINIA CURASSAVICA.

SURINAM POISON. See TEPHROSIA TOXICARIA.

SWEET WOOD, TIMBER. See NECTANDRA EXALTATA, AND N. LEUCANTHA.

SWEET WOOD, WHITE. See NECTANDRA LEUCANTHA.

SWEET CUP. See PASSIFLORA EDULIS, AND P. MALIFORMIS.

SWEET SOP. See ANONA SQUAMOSA.

SWIETENIA MAHAGONI, Linn.

MAHOGANY.

Native of Jamaica, Cuba, Bahamas, and Central America.

A lofty graceful tree; leaves pinnate; flowers small, greenish-yellow; seed-vessel opening by 5 valves from the base; and seeds flat, winged. *(Meliaceæ.)*

WOOD. "A well known and very durable wood, much used in general building and for furniture and ornamental work. Some of the

Jamaica mahogany is very fine. In the interior of the Island trees of considerable size are found." (Harrison.)

Bark astringent; used for diarrhœa by boiling an ounce of bruised bark in two pints of water down to one half. Bark of boughs, a good bitter, and febrifuge.

"Descending with the limestone to the sea coast on the shores of Westmoreland, St. Elizabeth. Elsewhere, further inland, at various elevations. Rare in the eastern districts. In the case of isolated trees up to 21 feet girth, with a spreading habit (Colbeck's, St. Catherine), but rarely exceeds four feet in diameter, and then only where preserved near dwelling-houses. In limestone forest, generally a small tree up to four feet growth, and 30 feet in height. Everywhere fruits in abundance. Timber employed in ornamental house and cabinet work. In the great houses of estates there are many specimens of beams and rafters of mahogany, very old and in good condition; but at the present time it is very seldom employed, only those trees being cut which are found on waste pastures and in forest near cultivation, and they rarely give than 10-inch planks. At no time has mahogany been largely exported from Jamaica, and recent trial shipments have been made at a loss. As a timber, the present stock is undoubtedly inferior to the Honduras (British and Spanish) varieties, having neither the ornamental grain and toughness of the one, nor the splendid dimensions acquired by the latter. With age it becomes of a good colour and is always a handsome wood. Formerly the wood from Jamaica was specially reputed for its mottled grain." (Hooper.)

SYMPHONIA GLOBULIFERA, Linn. fil.

Hog Gum.

Native of West Indies and tropical S. America.

A lofty tree, 90-100 feet high; leaves simple; flowers scarlet. (*Guttiferæ.*) (Generally known by the scientific name *Moronobea coccinea.*)

"In the interior hills. A lofty tree. Its resinous juice used in medicine, its timber in interior housework." (Hooper.)

TAMARIND. See Tamarindus indica.

TAMARIND, WILD. See Pithecolobium filicifolium.

TAMARINDUS INDICA, Linn.

Tamarind.

Cosmopolitan in the tropics.

A large tree, with pinnate leaves; flowers small, white or pale yellow with red veins; pod filled with pulp.

"Common on open plains. The heart-wood gives a handsome furniture timber. It grows to a large size on Goshen Common (St. Elizabeth)." (Hooper.)

FRUIT is prepared in the West Indies by removing the shell, placing alternate layers of fruit and sugar in a jar, and then pouring boiling syrup over them. In the East Indies, the shell is removed, and the fruit simply pressed together into a mass.

Tamarind pulp is laxative and refrigerant, and is also used to prepare a gargle for sore throat.

TEA. See CAMELLIA THEIFERA.

TEPHROSIA TOXICARIA, Pers.

SURINAM POISON.

Native of Trinidad, and tropical America.

A shrubby plant, 4 to 5 feet high; leaves pinnate; flowers white with a purplish tinge; pod 2 inches long. (*Leguminosæ.*)

Branches and leaves pounded, and thrown into river pools, stupefy fish. It is suggested that it might be used as a substitute for Digitalis.

TERMINALIA LATIFOLIA, Sw.

BROAD LEAF.

Native of Jamaica. A tree, 80 to 100 feet high; with alternate simple leaves, 6 to 12 inches long; flowers without petals; fruit $1\frac{1}{2}$ to 2 inches. (*Combretaceæ.*)

WOOD: "This tree grows to considerable size; it is not thought much of for building purposes, but as it splits readily, shingles and staves are often made from it." (Harrison.)

"Throughout the island. Very general on the limestone, and less represented at low elevations. Trees up to three feet diameter and 80 feet high, used in coopering and for shingles, but not prized for other uses." (Hooper.)

THEOBROMA CACAO, Linn.

CACAO. COCOA OR CHOCOLATE TREE.

Native of tropical America.

A small tree, with large, simple leaves, small pale-pink flowers, hanging in clusters from the branches and trunk; large pendulous fruit, full of seeds. (*Sterculiaceæ.*)

SEEDS used for the preparation of Cocoa and Chocolate.

THORN APPLE. See DATURA STRAMONIUM.

TOBACCO. See NICOTIANA TABACUM.

TORCH WOOD, MOUNTAIN. See AMYRIS BALSAMIFERA.

TOUS LES MOIS. See CANNA EDULIS.

TRAVELLER'S JOY. See CLEMATIS DIOICA.

TROPHIS AMERICANA, Linn.

RAMOON.

Native of Jamaica and Cuba.

A low tree; with milky juice; leaves simple, 8 to 4 inches long; flowers minute, several arranged together in spikes. (*Urticaceæ.*)

"Found along with, and on both sides of the zone occupied by, the breadnut. A similar growth to, but a smaller tree than the breadnut. Chiefly useful for the nutritious fodder it yields for cattle and horses." (Hooper.)

VANILLA PLANIFOLIA, Andr.

VANILLA.

Native of W. Indies, and tropical America. A climbing orchid, with pale yellowish green flowers, and a long 2-valved pod. (*Orchideæ.*)

PODS—forming Vanilla, gathered before they are quite ripe, and dried.

Vanilla used in perfumery, and for flavoring chocolate, liqueurs, &c.

VELVET LEAF. See CISSAMPELOS PAREIRA.

VITEX UMBROSA, Sw.

BOX WOOD, FIDDLE WOOD.

Native of Jamaica and Cuba.

A large tree; leaves compound, with 3 to 5 leaflets; flowers small with 2-lipped corolla, and yellow berries. (*Verbenaceæ.*)

WOOD. "Used for boards and framing purposes, is not a large tree but works up easily." (Harrison.)

VITIS CARIBÆA, DC.

WILD GRAPE, WATER WITHE.

Native of West Indies and tropical America.

A woody climber, with tendrils; leaves simple, with reddish-white down beneath; grapes small, purple. (*Ampelideæ.*)

The plant is known as "water-withe" from the circumstance that, in the early part of the year, the stem and large branches yield, when divided, about a pint of clear transparent fluid like water, which may be drunk.

FRUIT small, of the size of a currant, and has a rough acerb taste, recommending it for tarts. (Macfadyen.)

VITIS VINIFERA, Linn.

GRAPE VINE.

Native of South Europe and western Asia.

A shrubby plant climbing by means of tendrils. (*Ampelideæ.*)

FRUIT. Grapes are wholesome fruit, and the "grape cure" is recommended in Europe as a remedy in pulmonary diseases, etc. They are refrigerant, diuretic, and laxative.

WALNUT, INDIAN. See ALEURITES MOLUCCANA.

WATER LEMON. See PASSIFLORA LAURIFOLIA.

WATER WITHE. See VITIS CARIBÆA.

WINE PALM. See CARYOTA URENS.

WORM GRASS. See SPIGELIA ANTHELMIA.

XIMENIA AMERICANA, Linn.

MOUNTAIN PLUM.

Tropics. A tree; leaves simple; flowers white, in clusters; fruit yellow, size of a plum.

FRUIT.—Pulp, of a pleasant sub-acid taste, with a slight astringency.

YAM BEAN. See DOLICHOS TUBEROSUS.

YACCA. See PODOCARPUS CORIACEUS, & P. PURDIEANUS.

YELLOW SANDERS. See BUCIDA CAPITATA.

YOKE WOOD. See CATALPA LONGISSIMA.

ZANTHOXYLUM CLAVA-HERCULIS, Linn.

PRICKLY YELLOW.

Native of West Indies. A tree, 20-50 feet high, with pinnate leaves; flowers small, clustered in bunches. (*Rutaceæ.*)

WOOD.—"This wood is of a light yellow colour, saws readily straight, grows to 40 or 50 feet in height and 2 feet diameter at the butt, not considered durable for outside work." (Harrison.)

BARK.—Considered a powerful stimulant and sudorific, diuretic, and febrifuge. Used in rheumatism, paralysis of tongue, and as a febrifuge.

BARK OF ROOT, dried and powdered, is applied to sores. Infusion said to be antispasmodic.

"Throughout the island, especially common in the eastern districts up to 3,000 feet. A tall tree; with very cylindrical trunk, diameter up to three feet. The young growth cut for walking sticks on account of its rugged spines, which burst up through the bark. As a timber it is of a light yellow colour, and not considered durable when exposed in exterior work. Other species of this genus abounds, but do not attain any size." (Hooper.)

ZEA MAYS, Linn.

GREAT CORN. MAIZE. INDIAN CORN.

Native of America.

An annual grass, 4-10 feet high, with large leaves; flowers unisexual. *(Gramineæ.)*

Analysis of Maize yields in 100 parts, 54·37 starch, 8·83 nitrogenous substance, 4·50 fat, 2·70 gum and sugar, 15·77 cellulose, 12·16 water, and 1·67 ash. It contains less nitrogenous substance than wheat, and is therefore less nutritious, but it contains more fat than any other cereal.

ZINGIBER OFFICINALE, Rosc.

GINGER.

Native of tropical Asia.

A perennial herb, with a root-stock consisting of many roundish joints; leafy stems 3 or 4 feet high; leaves narrow, pointed with long sheathing stalks; flowering-stems 6 to 12 inches high; flowers small, yellow and purple, in a cone-like head. *(Zingiberaceæ.)*

ROOT-STOCK dried forms the ginger of commerce.

Ginger is used as a condiment, and the young shoots are preserved in syrup.

Medicinally, ginger possesses stimulant, aromatic, and carminative properties; and is of value in atonic dyspepsia. When chewed it acts as a sialagogue, and is employed in relaxed conditions of the uvula and tonsils. As a rubefacient, it relieves headache and toothache.

ZIZYPHUS CHLOROXYLON, Oliv.

COGWOOD.

Native of Jamaica.

A tree; leaves simple, alternate, 3-nerved.

WOOD. "A very hard and tough wood, twisted grain; is used for mill framings, cogwheels, &c." (Harrison.)

"Throughout the interior hills, notably St. Ann's. Grows to a large size, but most of the available timber has been destroyed, only small trees being found at present. The wood is dark, close-grained and heavy; used for coffee-mills and water-mills." (Hooper.)

www.ingramcontent.com/pod-product-compliance
Lightning Source LLC
Chambersburg PA
CBHW021959190326
41519CB00010B/1327

* 9 7 8 3 7 4 3 4 0 7 9 3 0 *